Landslide Disaster
Assessment and Monitoring

Landslide Disaster

Assessment and Monitoring

Ajay Mehta

RANDOM

RANDOM PUBLICATIONS

NEW DELHI - 110 002 (INDIA)

Landslide Disaster: Assessment and Monitoring

ISBN 978-93-51112-71-6

Published in 2014 in India by

RANDOM PUBLICATIONS

4376-A/4B, Gali Murari Lal, Ansari Road
New Delhi-110 002
Phone: +9111-43580356, 23289044
E-mail: randomexports@gmail.com; sales@randompublications.com; info@randompublications.com

Type Setting by: Friends Media, Delhi-110089
Digitally Printed at: Replika Press Pvt. Ltd.

Preface

Earthquake and landslide risk is a public safety issue that requires appropriate mitigation measures and means to protect citizens, property, infrastructure and the built cultural heritage. Mitigating this risk requires an integrated and coordinated action that embraces a wide range of organisations and disciplines. Earthquake and landslide risk is a public safety issue that requires appropriate mitigation measures and means to protect citizens, property, infrastructure and the built cultural heritage. Mitigating this risk requires an integrated and coordinated action that embraces a wide range of organisations and disciplines. Landslide risk assessment and disaster management are challenging problems in nowadays along with the progress of societal development towards mountainous area. A landslide disaster has a big influence to our social activity. In Japan, more than one million landslide places are located. And landslide disaster occurs every year at mountain area and residential region that located on steep terrain. To mitigate the landslide disaster, monitoring the entire landslide hazard site is necessary. But it needs a long time and a huge budget. Using satellite remote sensing technology is possible to apply as one solution of the large area monitoring. Natural disasters, whether of meteorological origin such as cyclones, floods, tornadoes and droughts or having geological nature such as earthquakes, volcanoes and landslide, are well known for their devastating impacts on human life, economy and environment.

Over recent decades, the people and the societies are becoming more vulnerable; although the frequency of natural events may be constant, human activities contribute to their increased intensity. Indeed, every year millions of people are affected by natural disasters globally and, only in the last decade, more than 80% of all disaster-related deaths were caused by natural hazards. The urban development, especially in recent decades, resulted in an increase of the assets at risk and unstable areas, often due to constant human intervention badly designed that led to instability also places previously considered "safe". Prevention is therefore essential to minimize the damages caused by landslides The objectives of the conducted research were to investigate the different techniques and to check their potentiality, in order to evaluate the most appropriate instrument for landslide hazard assessment in terms of better compromise between time to perform the analysis and expected results. The attempt is to evaluate which are the best methodologies to use according to the scenario, taking into consideration both reachable accuracies and time constraints. Careful considerations will be performed on strengths, weaknesses and limitations inherent to each methodology. The characteristics associated with geographic, or geospatial, information technologies facilitate the integration of scientific, social and economic data, opening up interesting possibilities for monitoring, assessment and change detection activities, thus enabling better informed interventions in human and natural systems. This is an important factor for the success of emergency operations and for developing valuable natural disaster preparedness, mitigation and prevention systems.

The publication is timely in view of the impressive development of the methods and techniques with their application to understanding the subject.

I thank all members of my team who have helped in the preparation of the book. My special thanks go to "Random Publications" who have published the book.

—Ajay Mehta

Contents

1

Introduction

Landslide

A landslide or landslip is a geological phenomenon which includes a wide range of ground movements, such as rockfalls, deep failure of slopes and shallow debris flows, which can occur in offshore, coastal and onshore environments. Although the action of gravity is the primary driving force for a landslide to occur, there are other contributing factors affecting the original slope stability. Typically, pre-conditional factors build up specific sub-surface conditions that make the area/slope prone to failure, whereas the actual landslide often requires a trigger before being released.

Causes of Landslides

The causes of landslides are usually related to instabilities in slopes. It is usually possible to identify one or more landslide causes and one landslide trigger. The difference between these two concepts is subtle but important. The landslide causes are the reasons that a landslide occurred in that location and at that time. Landslide causes are listed in the following table, and include geological factors, morphological factors, physical factors and factors associated with human activity.

Causes may be considered to be factors that made the slope vulnerable to failure, that predispose the slope to becoming unstable. The trigger is the single event that finally initiated the landslide. Thus, causes combine to make a slope vulnerable to failure, and the trigger finally initiates the movement. Landslides can have many causes but can only have one trigger as shown in the next figure.

Usually, it is relatively easy to determine the trigger after the landslide has occurred (although it is generally very difficult to determine the exact nature of landslide triggers ahead of a movement event).

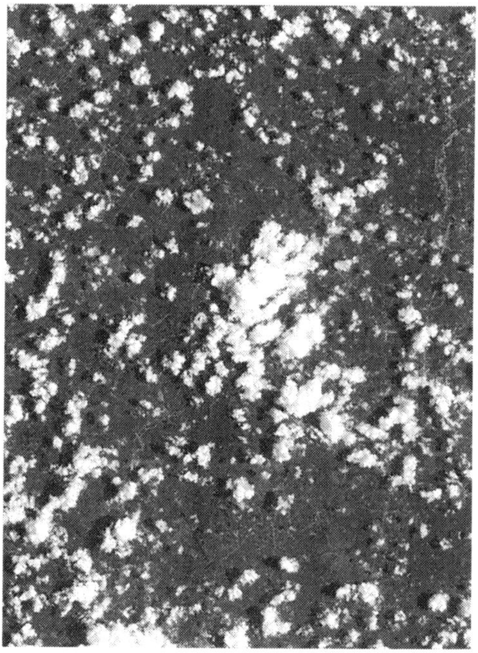

Figure: *Intense rain triggered widespread landslides in southern Thailand during the last week of March 2011.*

Occasionally, even after detailed investigations, no trigger can be determined - this was the case in the large Mount Cook landslide in New Zealand 1991. It is unclear as to whether the lack of a trigger in such cases is the result of some unknown process acting within the landslide, or whether there was in fact a trigger, but it cannot be determined. Perhaps this is because the trigger was in fact a slow but steady decrease in material strength associated with the weathering of the rock - at some point the material becomes so weak that failure must occur. Hence the trigger is the weathering process, but this is not detectable externally. In most cases we think of a trigger as an external stimulus that induces an immediate or near-immediate response in the slope, in this case in the form of the movement of the landslide. Generally this movement is induced either because the stresses in the slope are altered, perhaps by increasing shear stress or decreasing the effective normal stress, or by reducing the resistance to the movement perhaps by decreasing the shear strength of the materials within the landslide.

Causes of Landslides

Geological causes:

- Weathered Materials e.g heavy rainfall
- Sheared materials
- Jointed or fissured materials
- Adversely orientated discontinuities
- Permeability contrasts
- Material contrasts
- Rainfall and snow fall
- Earthquakes
- Working of machinery

Morphological Causes

- Slope angle
- Uplift
- Rebound
- Fluvial erosion
- Wave erosion
- Glacial erosion
- Erosion of lateral margins
- Subterranean erosion
- Slope loading
- Vegetation change
- Erosion

Physical Causes

- Intense rainfall
- Rapid snow melt
- Prolonged precipitation
- Rapid drawdown
- Earthquake
- Volcanic eruption
- Thawing
- Freeze-thaw
- Ground water changes

- Soil pore water pressure
- Surface runoff
- Seismic activity
- Soil erosion

Human Causes

- Excavation
- Loading
- Draw-down
- Land use change
- Water management
- Mining
- Quarrying
- Vibration
- Water leakage
- Deforestation
- Land use pattern
- Pollution

In the majority of cases the main trigger of landslides is heavy or prolonged rainfall. Generally this takes the form of either an exceptional short lived event, such as the passage of a tropical cyclone or even the rainfall associated with a particularly intense thunderstorm or of a long duration rainfall event with lower intensity, such as the cumulative effect of monsoon rainfall in South Asia. In the former case it is usually necessary to have very high rainfall intensities, whereas in the latter the intensity of rainfall may be only moderate - it is the duration and existing pore water pressure conditions that are important.

The importance of rainfall as a trigger for landslides cannot be underestimated. A global survey of landslide occurrence in the 12 months to the end of September 2003 revealed that there were 210 damaging landslide events worldwide. Of these, over 90% were triggered by heavy rainfall. One rainfall event for example in Sri Lanka in May 2003 triggered hundreds of landslides, killing 266 people and rendering over 300,000 people temporarily homeless. In July 2003 an intense rain band associated with the annual Asian monsoon tracked across central Nepal, triggering 14 fatal landslides that killed 85 people. The reinsurance company Swiss Re estimated that rainfall induced landslides associated with the 1997-1998 El Nino event triggered landslides along the west coast of North, Central and South America that resulted in over $5 billion in losses.

Finally, landslides triggered by Hurricane Mitch in 1998 killed an estimated 18,000 people in Honduras, Nicaragua, Guatemala and El Salvador. So why does rainfall trigger so many landslides? Principally this is because the rainfall drives an increase in pore water pressures within the soil.

The figure illustrates the forces acting on an unstable block on a slope. Movement is driven by shear stress, which is generated by the mass of the block acting under gravity down the slope.

Resistance to movement is the result of the normal load. When the slope fills with water, the fluid pressure provides the block with buoyancy, reducing the resistance to movement. In addition, in some cases fluid pressures can act down the slope as a result of groundwater flow to provide a hydraulic push to the landslide that further decreases the stability. Whilst the example given in Figures A and B is clearly an artificial situation, the mechanics are essentially as per a real landslide.

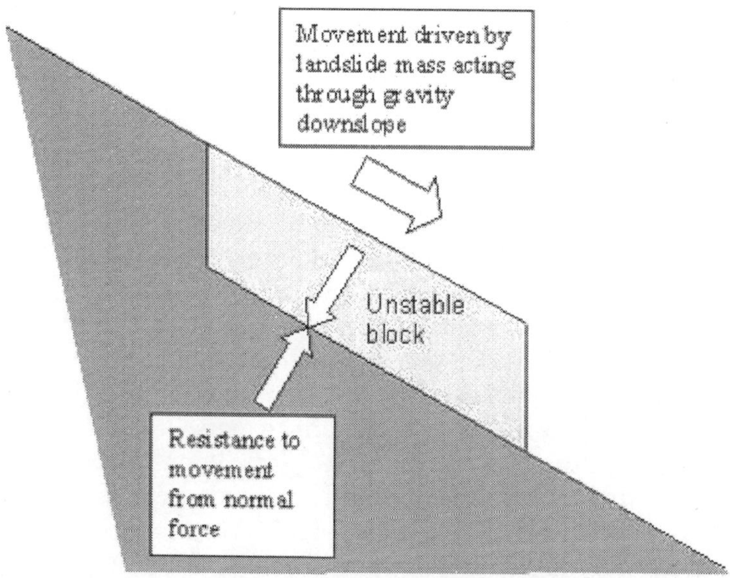

Figure: *A: Diagram illustrating the resistance to, and causes of, movement in a slope system consisting of an unstable block*

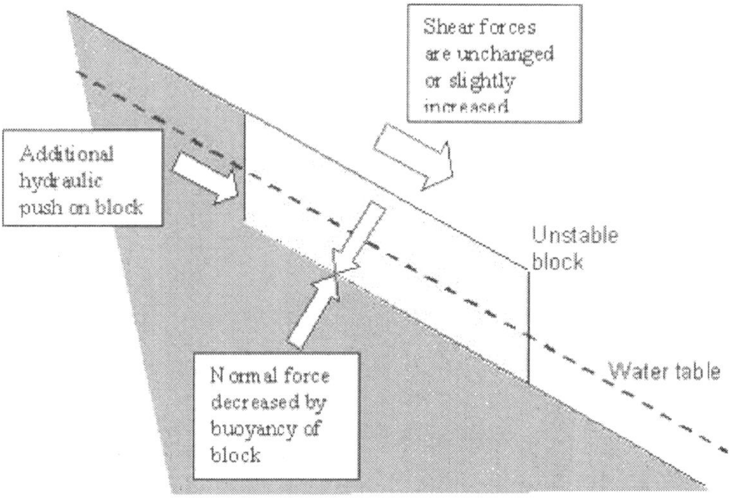

Figure: *B: Diagram illustrating the resistance to, and causes of, movement in a slope system consisting of an unstable block*

In some situations, the presence of high levels of fluid may destabilise the slope through other mechanisms, such as:

- Fluidization of debris from earlier events to form debris flows;
- Loss of suction forces in silty materials, leading to generally shallow failures (this may be an important mechanism in residual soils in tropical areas following deforestation);
- Undercutting of the toe of the slope through river erosion.

Considerable efforts have been made to understand the triggers for landsliding in natural systems, with quite variable results. For example, working in Puerto Rico, Larsen and Simon found that storms with a total precipitation of 100–200 mm, about 14 mm of rain per hour for several hours, or 2–3 mm of rain per hour for about 100 hours can trigger landslides in that environment. Rafi Ahmad, working in Jamaica, found that for rainfall of short duration (about 1 hour) intensities of greater than 36 mm/h were required to trigger landslides. On the other hand, for long rainfall durations, low average intensities of about 3 mm/h appeared to be sufficient to cause landsliding as the storm duration approached approximately 100 hours. Corominas and Moya (1999) found that the following thresholds exist for the upper basin of the Llobregat River, Eastern Pyrenees area. Without antecedent rainfall, high intensity and short duration rains triggered debris flows and shallow slides developed in colluvium and weathered rocks. A rainfall threshold of around 190 mm in 24 h initiated failures whereas more than 300 mm in 24-48 h were needed to cause widespread shallow landsliding. With antecedent rain, moderate intensity precipitation of at least 40 mm in 24 h reactivated mudslides and both rotational and translational slides affecting clayey and silty-clayey formations. In this case, several weeks and 200 mm of precipitation were needed to cause landslide reactivation. A similar approach is reported by Brand et al. (1988) for Hong Kong, who found that if the 24 hour antecedent rainfall exceeded 200 mm then the rainfall threshold for a large landslide event was 70 mm hr-1. Finally, Caine (1980) established a worldwide threshold:

I = 14.82 D - 0.39 *where: I is the rainfall intensity (mm h-1), D is duration of rainfall (h)*

This threshold applies over time periods of 10 minutes to 10 days. It is possible to modify the formula to take into consideration areas with high mean annual precipitations by considering the proportion of mean annual precipitation represented by any individual event. Other techniques can be used to try to understand rainfall triggers, including:

- Actual rainfall techniques, in which measurements of rainfall are adjusted for potential evapotranspiration and then correlated with landslide movement events
- Hydrogeological balance approaches, in which pore water pressure response to rainfall is used to understand the conditions under which failures are initiated
- Coupled rainfall - stability analysis methods, in which pore water pressure response models are coupled to slope stability models to try to understand the complexity of the system
- Numerical slope modelling, in which finite element (or similar) models are used to try to understand the interactions of all relevant processes

Snowmelt

In many cold mountain areas, snowmelt can be a key mechanism by which landslide initiation can occur. This can be especially significant when sudden increases in temperature lead to rapid melting of the snow pack. This water can then infiltrate into the ground, which may have impermeable layers below the surface due to still-frozen soil or rock, leading to rapid increases in pore water pressure, and resultant landslide activity. This effect can be especially serious when the warmer weather is accompanied by precipitation, which both adds to the groundwater and accelerates the rate of thawing.

Water-Level Change

Rapid changes in the groundwater level along a slope can also trigger landslides. This is often the case where a slope is adjacent to a water body or a river. When the water level adjacent to the slope falls rapidly the groundwater level frequently cannot dissipate quickly enough, leaving an artificially high water table. This subjects the slope to higher than normal shear stresses, leading to potential instability. This is probably the most important mechanism by which river bank materials fail, being significant after a flood as the river level is declining (i.e. on the falling limb of the hydrograph) as shown in the following figures.

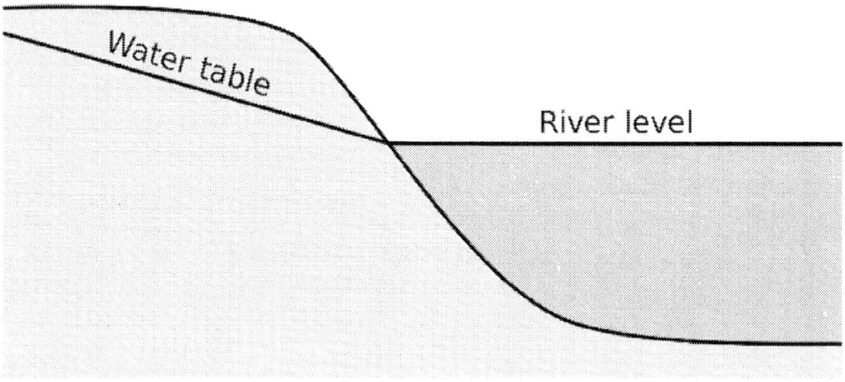

Figure: *Groundwater conditions when the river level is stable*

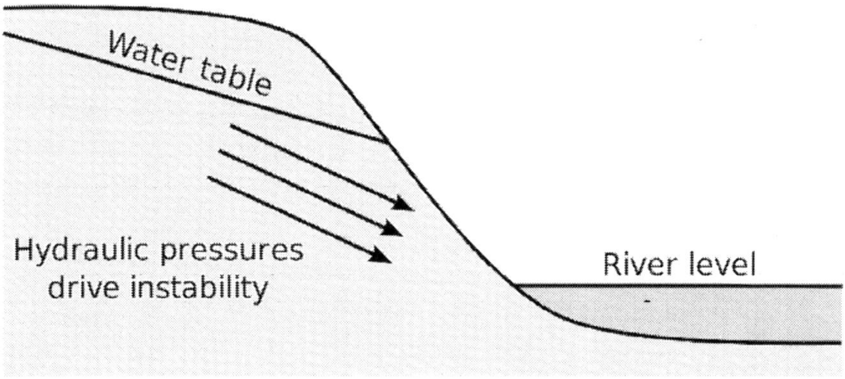

Figure: *Groundwater conditions on the falling limb of the hydrograph. If the fall in river levels is sufficiently rapid then the high water levels in the slope can provide a hydraulic push that destabilises the slope, sometimes triggering bank collapse*

It can also be significant in coastal areas when sea level falls after a storm tide, or when the water level of a reservoir or even a natural lake rapidly falls. The most famous example of this is the Vajont failure, when a rapid decline in lake level contributed to the occurrence of a landslide that killed over 2000 people.

Rivers

In some cases, failures are triggered as a result of undercutting of the slope by a river, especially during a flood. This undercutting serves both to increase the gradient of the slope, reducing stability, and to remove toe weighting, which also decreases stability. For example, in Nepal this process is often seen after a glacial lake outburst flood, when toe erosion occurs along the channel. Immediately after the passage of flood waves extensive landsliding often occurs. This instability can continue to occur for a long time afterwards, especially during subsequent periods of heavy rain and flood events.

Seismicity

The second major factor in the triggering of landslides is seismicity. Landslides occur during earthquakes as a result of two separate but interconnected processes: seismic shaking and pore water pressure generation.

Seismic Shaking

The passage of the earthquake waves through the rock and soil produces a complex set of accelerations that effectively act to change the gravitational load on the slope. So, for example, vertical accelerations successively increase and decrease the normal load acting on the slope. Similarly, horizontal accelerations induce a shearing force due to the inertia of the landslide mass during the accelerations. These processes are complex, but can be sufficient to induce failure of the slope. These processes can be much more serious in mountainous areas in which the seismic waves interact with the terrain to produce increases in the magnitude of the ground accelerations. This process is termed 'topographic amplification'. The maximum acceleration is usually seen at the crest of the slope or along the ridge line, meaning that it is a characteristic of seismically triggered landslides that they extend to the top of the slope.

Liquefaction

The passage of the earthquake waves through a granular material such as a soil can induce a process termed liquefaction, in which the shaking causes a reduction in the pore space of the material. This densification drives up the pore pressure in the material. In some cases this can change a granular material into what is effectively a liquid, generating 'flow slides' that can be rapid and thus very damaging. Alternatively, the increase in pore pressure can reduce the normal stress in the slope, allowing the activation of translational and rotational failures.

The Nature of Seismically-Triggered Landslides

For the main part seismically generated landslides usually do not differ in their morphology and internal processes from those generated under non-seismic conditions. However, they tend to be more widespread and sudden. The most abundant types of earthquake-induced landslides are rock falls and slides of rock fragments that form on steep slopes. However, almost every other type of landslide is possible, including highly disaggregated and fast-moving falls; more coherent and slower-moving slumps, block slides, and earth slides; and lateral spreads and flows that involve partly to completely liquefied material (Keefer, 1999). Rock falls, disrupted rock slides, and disrupted slides of earth and debris are the most abundant types of earthquake-induced landslides, whereas earth flows, debris flows, and avalanches of rock, earth, or debris typically transport material the farthest. There is one type of landslide that is essential uniquely limited to earthquakes - liquefaction failure, which can cause fissuring or subsidence of the ground. Liquefaction involves the temporary loss of strength of sands and silts which behave as viscous fluids rather than as soils. This can have devastating effects during large earthquakes.

Volcanic Activity

Some of the largest and most destructive landslides known have been associated with volcanoes. These can occur either in association with the eruption of the volcano itself, or as a result of mobilisation of the very weak deposits that are formed as a consequence of volcanic activity. Essentially, there are two main types of volcanic landslide: lahars and debris avalanches, the largest of which are sometimes termed flank collapses. An example of a lahar was seen at Mount St Helens during its catastrophic eruption on May 18, 1980. Failures on volcanic flanks themselves are also common. For example, a part of the side of Casita Volcano in Nicaragua collapsed on October 30, 1998 during the heavy precipitation associated with the passage of Hurricane Mitch. Debris from the initial small failure eroded older deposits from the volcano and incorporated additional water and wet sediment from along its path, increasing in volume about ninefold. The lahar killed more than 2,000 people as it swept over the towns of El Porvenir and

Rolando Rodriguez at the base of the mountain. Debris avalanches commonly occur at the same time as an eruption, but occasionally they may be triggered by other factors such as a seismic shock or heavy rainfall. They are particularly common on strato volcanoes, which can be massively destructive due to their large size. The most famous debris avalanche occurred at Mount St Helens during the massive eruption in 1980. On May 18, 1980, at 8:32 a.m. local time, a magnitude 5.1 earthquake shook Mount St. Helens. The bulge and surrounding area slid away in a gigantic rockslide and debris avalanche, releasing pressure, and triggering a major pumice and ash eruption of the volcano. The debris avalanche had a volume of about 1 km^3 (0.24 cu mi), traveled at 50 to 80 m/s (110 to 180 mph), and covered an area of 62 km^2 (24 sq mi), killing 57 people.

Landslide Classification

There have been known various classifications of landslides and other types of mass wasting.

For example, the *McGraw-Hill Encyclopedia of Science and Technology* distinguishes the following types of landslides:

- fall (by undercutting)
- fall (by toppling)
- slump
- June 2011
- rockslide
- earthflow
- rockslide that develops into rock avalanche

The Used Classification Factors

Various scientific disciplines have developed taxonomic classification systems to describe natural phenomena or individuals, like for example, plants or animals. These systems are based on specific characteristics like shape of organs or nature of reproduction. Differently, in landslide classification, there are great difficulties due to the fact that phenomena are not perfectly repeatable; usually being characterised by different causes, movements and morphology, and involving genetically different material. For this reason, landslide classifications are based on different discriminating factors, sometimes very subjective. In the following factors are discussed by dividing them into two groups: the first one is made up of the criteria utilised in the most widespread classification systems that can generally be easily determined. The second one is formed by those factors that have been utilised in some classifications and can be useful in descriptions

Type of Movement

This is the most important criteria, even if uncertainties and difficulties can arise in the identification of movements, being the mechanisms of some landslides often particularly complex. The main movements are falls, slides and flows, but usually topples, lateral spreading and complex movements are added to these. They can be deadly.

Involved Material

Rock, earth and debris are the terms generally used to distinguish the materials involved in the landslide process. For example, the distinction between earth and debris is usually made by comparing the percentage of coarse grain size fractions. If the weight of the particles with a diametre greater than 2 mm is less than 20%, the material will be defined as earth; in the opposite case, it is debris.

Activity

The classification of a landslide based on its activity is particularly relevant in the evaluation of future events. The recommendations of the WP/WLI (1993) define the concept of activity with reference to the spatial and temporal conditions, defining the state, the distribution and the style.

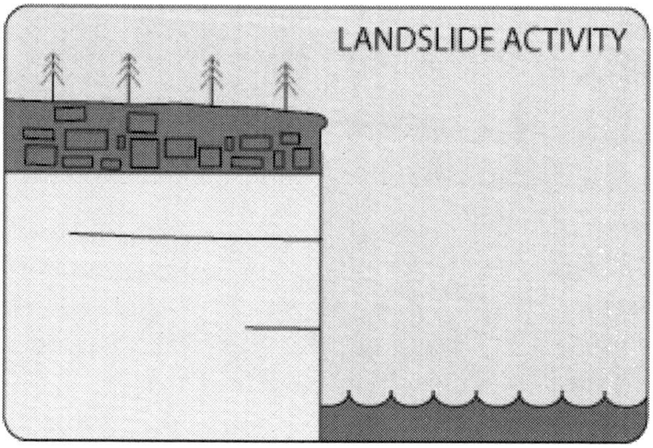

The first term describes the information regarding the time in which the movement took place, permitting information to be available on future evolution, the second term describes, in a general way, where the landslide is moving and the third term indicates how it is moving.

Movement Velocity

This factor has a great importance in the hazard evaluation. A velocity range is connected to the different type of landslides, on the basis of observation of case history or site observations.

The Age of the Movement

Landslide dating is an interesting topic in the evaluation of hazard. The knowledge of the Landslide frequency is a fundamental element for any kind of probabilistic evaluation. Furthermore, the evaluation of the age of the landslide permits to correlate the trigger to specific conditions, as earthquakes or periods of intense rains. It should be noted that, it is possible that phenomena could be occurred in past geological times, under specific environmental conditions which no longer act as agents today. For example, in some Alpine areas, landslides of the Pleistocene age are connected with particular tectonic, geomorphological and climatic conditions.

B2) Geological conditions This represent a fundamental factor of the morphological evolution of a slope. Bedding attitude and the presence of discontinuities or faults control the slopemorphogenesis.

Morphological Characteristics

As the landslide is a geological volume with a hidden side, morphological characteristics are extremely important in the reconstruction of the technical model.

Geographical Location

This criterion describe, in a general way, the location of landsides in the physiographic context of the area. Some authors have therefore identified landslides according to their geographical position so that it is possible to describe "alpine landslides", "landslides in plains", "hilly landslides" or "cliff landslides". As a consequence, specific morphological contexts are referred characterised by slope evolution processes.

Topographical Criteria

With these criteria, landslides can be identified with a system similar to that of the denomination of formations. Consequently, it is possible to describe a landslide using the name of a site. In particular, the name will be that of the locality where the landslide happened with a specific characteristic type.

Type of Climate

These criteria give particular importance to climate in the genesis of phenomena for which similar geological conditions can, in different climatic conditions, lead to totally different morphological evolution. As a consequence, in the description of a landslide, it can be interesting to understand in what type of climate the event occurred.

Causes of the Movements

In the evaluation of landslide susceptibility, causes of the triggers is an important step. Terzaghi describes causes as "internal" and "external" referring to modifications in the conditions of the stability of the bodies. Whilst the internal causes induce modifications in the material itself which decrease its resistance to shear stress, the external causes generally induce an increase of shear stress, so that block or bodies are no longer stable. The triggering causes induce the movement of the mass. Predisposition to movement due to control factors is determining in landslide evolution. Structural and geological factors, as already described, can determine the development of the movement, inducing the presence of mass in kinematic freedom.

Types and Classification

In the following table shows a schematic landslide classification adopting the classification of Varnes 1978 and taking into account the modifications made by Cruden and Varnes, in 1996. Some integration have been made by using the definitions of Hutchinson (1988) and Hungr et al. 2001.

Type of movement			***Type of material***		
			Bedrock	***Engineering soils***	
				Predominantly fine	***Predominantly coarse***
Falls			Rockfall	Earth fall	Debris fall
Topples			Rock topple	Earth topple	Debris topple
Slides	Rotational		Rock slump	Earth slump	Debris slump
	Translational	Few units	Rock block slide	Earth block slide	Debris block slide
		Many units	Rock slide	Earth slide	Debris slide
Lateral spreads			Rock spread	Earth spread	Debris spread
Flows			Rock flow	Earth flow	Debris flow
			Rock avalanche		Debris avalanche
			(Deep creep)	(Soil creep)	
Complex and compound			Combination in time and/or space of two or more principal types of movement		

Falls

Description: *"A fall starts with the detachment of soil or rock from a steep slope along a surface on which little or no shear* *displacement takes place. The material then descends mainly through the air by falling, bouncing, or rolling"* (Varnes, 1996).

Secondary Falls: *"Secondary falls involves rock bodies already physically detached from* cliff *and merely lodged upon it"* (Hutchinson, 1988)

Speed: *from very to extremely rapid*

Type of Slope: *slope angle 45-90 degrees*

Control Factor: *Discontinuities*

Causes: *Vibration, undercutting, differential* weathering, *excavation, or stream erosion*

Topples

Description: *"Topples is the forward* rotation *out of the slope of mass of soil or rock about a point or* axis *below the* centre of gravity *of the displaced mass. Toppling is sometimes driven by gravity exerted by material upslope of the displaced mass and sometimes by water or ice in cracks in the mass"* (Varnes, 1996)

Speed: *extremely slow to extremely rapid*

Type of Slope: *slope angle 45-90 degrees*

Control Factor: *Discontinuities, lithostratigraphy*

Causes: *Vibration, undercutting, differential weathering, excavation, or stream erosion*

Slides

"A slide is a downslope movement of soil or rock mass occurring dominantly on the surface of rupture or on relatively thin zones of intense shear strain." (Varnes, 1996)

Translational Slide

Description: *"In translational slides the mass displaces along a planar or undulating surface of rupture, sliding out over the original ground surface."* (Varnes, 1996)

Speed: *extremely slow to extremely rapid (>5 m/s)*

Type of Slope: *slope angle 45-90 degrees*

Control Factor: *Discontinuities, geological setting*

Causes: *Vibration, undercutting, differential weathering, excavation, or stream erosion*

Rotational Slides

Description: *"Rotational slides move along a surface of rupture that is curved and concave"* (Varnes, 1996)

Speed: *extremely slow to extremely rapid*

Type of Slope: *slope angle 45-90 degrees*

Control Factor: *morphology and lithology*

Causes: *Vibration, undercutting, differential weathering, excavation, or stream erosion*

Spreads

Description: *"Spread is defined as an extension of a cohesive soil or rock mass combined with a general subsidence of the fractured mass of cohesive material into softer underlying material." (Varnes,* 1996)." In *spread, the dominant mode of movement is lateral extension accommodated by shear or tensile fractures"* (Varnes, 1978)

Speed: *extremely slow to extremely rapid (>5 m/s)*

Type of Slope: *angle 45-90 degrees*

Control Factor: *Discontinuities, lithostratigraphy*

Causes: *Vibration, undercutting, differential weathering, excavation, or stream erosion*

Flows

A flow is a spatially continuous movement in which surfaces of shear are short-lived, closely spaced, and usually not preserved. The distribution of velocities in the displacing mass resembles that in a viscous liquid. The lower boundary of displaced mass may be a surface along which appreciable differential movement has taken place or a thick zone of distributed shear (Cruden&Varnes, 1996)

Flows in Rock

Rock Flow: *Description: "Flow movements in bedrock include deformations that are distributed among many large or small fractures, or even microfracture, without concentration of displacement along a through-going fracture"* (Varnes, 1978)

Speed: *extremely slow*

Type of Slope: *angle 45-90 degrees*

Causes: *Vibration, undercutting, differential weathering, excavation, or stream erosion*

Rock avalanche (Sturzstrom): *Description: "Extremely rapid, massive, flow-like motion of fragmented rock from a large rock slide or rock fall"* (Hungr, 2001)

Speed: *extremely rapid*

Type of Slope: *angle 45-90 degrees*

Control Factor: *Discontinuities, lithostratigraphy*

Causes: *Vibration, undercutting, differential weathering, excavation or stream erosion.*

Flows in Soil

Debris flow: *Description: "Debris flow is a very rapid to extremely rapid flow of saturated non-plastic debris in a steep channel"* (Hungr et al.,2001)

Speed: *very rapid to extremely rapid (>5 m/s)*

Type of Slope: *angle 20-45 degrees*

Control Factor: *torrentsediments, water flows*

Causes: *High intensity rainfall*

Debris avalanche

Description: *"Debris avalanche is a very rapid to extremely rapid shallow flow of partially or fully saturated debris on a steep slope, without confinement in an established channel."* (Hungr et al., 2001)

Speed: *very rapid to extremely rapid (>5 m/s)*

Type of Slope: *angle 20-45 degrees*

Control Factor: *morphology, regolith*

Causes: *High intensity rainfalls*

Figure: *Location: Castel franci - Italy Landslide type: Earth flow*

Earth flow:

Description: *"Earth flow is a rapid or slower, intermittent flow-like movement of plastic, clayey earth."* (Hungr et al.,2001)

Speed: *slow to rapid (>1,8 m/h)*

Type of Slope: *slope angle 5-25 degrees*

Control Factor: *lithology*

Mudflow

Description: *"Mudflow is a very rapid to extremely rapid flow of saturated plastic debris in a channel, involving significantly greater water content relative to the source material (Plasticity index> 5%)."* (Hungr et al.,2001)

Speed: *very rapid to extremely rapid (>5 m/s)*

Type of Slope: *angle 20-45 degrees*

Control Factor: *torrentsediments, water flows*

Causes: *High intensity rainfall*

Complex Movement

Description: *Complex movement is a combination of falls, topples, slides, spreads and flows*

Landslide Prediction Mapping

Figure: *Ferguson Slide on California State Route 140 in June 2006*

Landslide hazard analysis and mapping can provide useful information for catastrophic loss reduction, and assist in the development of guidelines for sustainable land use planning. The analysis is used to identify the factors that are related to landslides, estimate the relative contribution of factors causing slope failures, establish a relation between the factors and landslides, and to predict the landslide hazard in the future based on such a relationship. The factors that have been used for landslide hazard analysis can usually be grouped into geomorphology, geology, land use/land cover, and hydrogeology. Since many factors are considered for landslide hazard mapping, GIS is an appropriate tool because it has functions of collection, storage, manipulation, display, and analysis of large amounts of spatially referenced data which can be handled fast and effectively. Remote sensing techniques are also highly employed for landslide hazard assessment and analysis. Before and after aerial photographs and satellite imagery are used to gather landslide characteristics, like distribution and classification, and factors like slope, lithology, and land use/land cover to be used to help predict future events. Before and after imagery also helps to reveal how the landscape changed after an event, what may have triggered the landslide, and shows the process of regeneration and recovery.

Figure: *Trackside rock slide detector on the UPRR Sierra grade near Colfax, CA*

Using satellite imagery in combination with GIS and on-the-ground studies, it is possible to generate maps of likely occurrences of future landslides. Such maps should show the locations of previous events as well as clearly indicate the probable locations of future events. In general, to predict landslides, one must assume that their occurrence is determined by certain geologic factors, and that future landslides will occur under the same conditions as past events. Therefore, it is necessary to establish a relationship between the geomorphologic conditions in which the past events took place and the expected future conditions.

Natural disasters are a dramatic example of people living in conflict with the environment. Early predictions and warnings are essential for the reduction of property damage and loss of life. Because landslides occur frequently and can represent some of the most destructive forces on earth, it is imperative to have a good understanding as to what causes them and how people can either help prevent them from occurring or simply avoid them when they do occur. Sustainable land management and development is an essential key to reducing the negative impacts felt by landslides.

GIS offers a superior method for landslide analysis because it allows one to capture, store, manipulate, analyze, and display large amounts of data quickly and effectively. Because so many variables are involved, it is important to be able to overlay the many layers of data to develop a full and accurate portrayal of what is taking place on the Earth's surface. Researchers need to know which variables are the most important factors that trigger landslides in any given location. Using GIS, extremely detailed maps can be generated to show past events and likely future events which have the potential to save lives, property, and money.

Prehistoric Landslides

- Landslide which moved Heart Mountain to its current location, the largest ever discovered on land. In the 48 million years since the slide occurred, erosion has removed most of the portion of the slide.
- Flims Rockslide, ca. 12 km^3 (2.9 cu mi), Switzerland, some 10000 years ago in post-glacial Pleistocene/Holocene, the largest so far described in the alps and on dry land that can be easily identified in a modestly eroded state.
- The landslide around 200BC which formed Lake Waikaremoana on the North Island of New Zealand, where a large block of the Ngamoko Range slid and dammed a gorge of Waikaretaheke River, forming a natural reservoir up to 248 metres deep.
- Cheekye Fan, British Columbia, Canada, ca. 25 km^2 (9.7 sq mi), Late Pleistocene in age.

Prehistoric Submarine Landslides

- The Storegga Slide, Norway, ca. 3,500 km^3 (840 cu mi), ca. 8,000 years ago, a catastrophic impact on the contemporary coastal Mesolithic population
- The Agulhas slide, ca. 20,000 km^3 (4,800 cu mi), off South Africa, post-Pliocene in age, the largest so far described
- The Ruatoria Debris Avalanche, off North Island New Zealand, ca. 3,000 km^3 in volume, 170,000 years ago.
- Catastrophic debris avalanches have been common on the submerged flanks of ocean island volcanos such as the Hawaiian Islands and the Cape Verde Islands.

Historical Landslides

- The Goldau on September 2, 1806
- The Cap DiamantQuébec rockslide on September 19, 1889
- Frank Slide, Turtle Mountain, Alberta, Canada, on 29 April 1903
- Khait landslide, Khait, Tajikistan, Soviet Union, on July 10, 1949
- Monte Toc landslide (260 millions cubic metres) falling into the Vajont Dam basin in Italy, causing a mega tsunami and about 2000 deaths, on October 9, 1963
- Hope Slide landslide (46 million cubic metres) near Hope, British Columbia on January 9, 1965.
- The 1966 Aberfan disaster
- Tuve landslide in Gothenburg, Sweden on November 30, 1977.
- The 1979 Abbotsford landslip, Dunedin, New Zealand on August 8, 1979.
- Val Pola landslide during Valtellina disaster (1987) Italy
- Thredbo landslide, Australia on 30 July 1997, destroyed hostel.
- Vargas mudslides, due to heavy rains in Vargas State, Venezuela, on December, 1999, causing tens of thousands of deaths.
- 2007 Chittagong mudslide, in Chittagong, Bangladesh, on June 11, 2007.
- 2008 Cairo landslide on September 6, 2008.

- The 2010 Uganda landslide caused over 100 deaths following heavy rain in Bududa region.
- Zhouqu county mudslide in Gansu, China on August 8, 2010.
- Devil's Slide, an ongoing landslide in San Mateo County, California
- 2011 Rio de Janeiro landslide in Rio de Janeiro, Brazil on January 11, 2011, causing 610 deaths.

Extraterrestrial Landslides

Evidence of past landslides has been detected on many bodies in the solar system, but since most observations are made by probes that only observe for a limited time and most bodies in the solar system appear to be geologically inactive not many landslides are known to have happened in recent times. Both Venus and Mars have been subject to long-term mapping by orbiting satellites, and examples of landslides have been observed on both.

Deformation Monitoring

Deformation monitoring (also referred to as Deformation survey) is the systematic measurement and tracking of the alteration in the shape or dimensions of an object as a result of stresses induced by applied loads. Deformation monitoring is a major component of logging measured values that may be used to for further computation, deformation analysis, predictive maintenance and alarming.

Deformation monitoring is primarily related to the field of applied surveying, but may be also related to the civil engineering, mechanical engineering, plant construction, soil and rock stability mechanics.

The causes for deformation monitoring are changes in the bedrock, increase or decrease of weight, changes of the material properties or outside influences.

The used measuring devices (1) for a deformation monitoring depend on the application (2), the chosen method (3) and the required regularity (4).

Measuring Devices

Measuring devices (or sensors) can be sorted in two main groups, geodetic and geotechnical sensors. Both measuring devices can be seamlessly combined in modern deformation monitoring.

- Geodetic measuring devices measure georeferenced displacements or movements in one, two or three dimensions.

It includes the use of instruments such as total stations, levels and global navigation satellite system receivers.

- Geotechnical measuring devices measure non-georeferenced displacements or movements and related environmental effects or conditions. It includes the use of instruments such as extensometres, piezometres, rain gauges, thermometres, barometres, tilt metres, accelerometres, seismometres etc.
- Other techniques e.g. radar measuring devices.

Application

Deformation monitoring can be required for the following applications:

- Dams
- Roads
- Tunnels
- Bridges and Viaducts
- High-rise and historical buildings
- Foundations
- Construction sites
- Mining
- Landslide and Volcanoes Slopes
- Settlement areas
- Earthquake areas

Methods

Deformation monitoring can be made manually or automatically.

- Manual deformation monitoring is the operation of sensors or instruments by hand for the purpose of deformation monitoring.
- An automatic deformation monitoring system is a group of interacting, interrelated, or interdependent software and hardware elements forming a complex whole for deformation monitoring that, once set up, does not require human input to function.

Note that deformation analysis and interpretation of the data collected by the monitoring system is not included in this definition. An automatic monitoring system may be used for periodic or continuous monitoring.

Regularity and Scheduling

The monitoring regularity and time interval of the measurements must be considered depending on the application and object to be monitored. Objects can undergo both rapid, high frequency movement and slow, gradual movement. For example, a bridge might oscillates with a period of a few seconds due to the influence of traffic and wind and also be shifting gradually due to tectonic changes.

- Regularity: ranges from a days, weeks or years for manual monitoring and continuous for automatic monitoring systems.
- Measurement interval: ranges from fractions of a second to hours.

Risk Management

Deformation monitoring systems provide a proactive control of a hazard related to possible change or failure of a structure. Policyholders can reduce risk exposure before and during construction and throughout the lifecycle of the structure and hence decrease the insurance premium.

2

Landslide Mitigation

Landslide mitigation refers to lessening the effect of landslides by constructing various man made projects on slopes which are vulnerable to landslides. Landslides can be triggered by many often concomitant causes. In addition to shallow erosion or reduction of shear strength caused by seasonal rainfall, causes triggered by anthropic activities such as adding excessive weight above the slope, digging at mid-slope or at the foot of the slope, can also be included. However, often individual phenomena join together to generate instability, also after some time has elapsed, which, other than in well-instrumented limited areas, do not allow a reconstruction of the evolution of the occurred landslide. It is therefore pointless, for the purpose of planning landslide hazard mitigation measures, to classify the work as a function of the phenomenon or of more important phenomena, renouncing any attempt to precisely describe all the causes or the conditions which, at different times, contribute to the occurrence of the landslide. Therefore, slope stabilisation methods in rock or in earth, can be collocated into three types of measure:

- Geometric methods, in which the geometry of the hillside is changed (in general the slope);
- Hydrogeological methods, in which an attempt is made to lower the groundwater level or to reduce the water content of the material;
- Chemical and mechanical methods, in which attempts are made to increase the shear strength of the unstable mass or to introduce active external forces (e.g. anchors, rock or ground nailing) or passive (e.g. structural wells, piles or reinforced ground) to contrast the destabilising forces.

The different type of material conditions the engineering solution adopted, although It always comes back, in principle, to the previously introduced classification.

Rock Slopes

Reinforcement Measures

Reinforcement measures generally consist of the introduction of metal elements whose purpose is to increase the shear strength of the rock and to reduce the stress release created, for example, following cutting. Reinforcement measures are made up of metal rock nails or anchors. Anchorage can be classified as active anchorage, in the case in which they are subjected to pretensioning, and passive anchorage. Passive anchorage can be used both to nail single unstable blocks and to reinforce large portions of rock. They can also be used as the pre-reinforcement elements of a scarp to be re-profiled in order to limit hillside decompression associated with cutting. In an anchorage are defined:

- the header: the set of elements (anchor plate, blocking device, etc.) with the purpose of transmitting the traction strength of the anchor to the anchored structure or to the rock,
- the reinforcement: part of the anchor, concreted and otherwise, placed under traction; can be constituted by a metal rod, a metal cable, a strand, etc.;
- the length of the foundation: the deepest portion of the anchor, fixed to the rock with chemical bonds or mechanical devices, which transfer the load to the rock itself,
- the free length: the non-concreted length.

When the anchorage acts over a short length it is defined as a bolt. It is therefore a specific type of anchorage, not structurally connected to the free length, made up of an element resistant to traction, normally a steel bar of less than 12 m, protected against corrosion by a concrete sheath. As far as the anchorage device to the ground is concerned, it can be chemical, or use mechanical expansion or concreting.

In the first case, some polyester resin cartridges are placed in the perforation to fill the ring space around the end part of the bolt. The main advantage of this type of anchorage lies in its simplicity and in the speed of installation. The main disadvantage is in its limited strength. In the second case, the anchorage is composed of steel wedges driven into the sides of the hole. The advantage of this type of anchorage lies in the speed of installation and in the fact that the tensioning can be achieved in the instant the anchorage is put to work. The main disadvantage with this type of anchorage is that it can only be used with hard rock. Moreover, the maximum traction force is limited.

In the third case, the anchorage is obtained by concreting the whole metal bar. This is the most-used method since the materials are cheap and installation is simple. Injected concrete mixes can be used in many different rocks and grounds; moreover, the concrete sheath protects the bar from corrosion. The concrete mixture is generally made up of water and cement in the ratio $W/C = 0.40\text{-}0.45$, since in this way a sufficiently fluid mixture is obtained to allow pumping into the hole, while at the same time, when set, providing high mechanical strength. As far as the working mechanism of a rock nail is concerned, the strains of the rock induce a stress state in the nail composed of shear and traction stress, due to the roughness of the joints, to their opening and to the direction of the nail, generally non-orthogonal to the joint itself. The execution phases of setting up the nail provides for:

- formation of any header niche and perforation;
- setting up of a reinforcement bar (e.g. a 4–6 m long FeB44k bar);
- concrete injection of the bar;
- sealing of the header or of the top part of the hole.

It is anyway opportune to close up and cement any cracks in the rock to prevent pressure caused by water during the freeze-thaw cycles from producing progressive breaking in the reinforcement system set up. To this purpose a procedure is provided for of:

- cleaning out and washing of the cracks;
- plastering of the crack;
- predisposition of the injection tubes at suitable inter-axes, parallel to the crack, through which the concrete mix is injected;
- sequential injection of the mixture from bottom to top and at low pressure (1-3 atm.) until refusal or until no flow back of the mixture is noted from the tubes placed higher up.

The injection mixtures will have approximately the following composition:

- cement 10 kg;
- water 65 l
- fluidity and anti-shrinkage additive or bentonite 1-5 kg.

Shotcrete

As defined by the American Concrete Institute, shotcrete is mortar or concrete conveyed through a hose and pneumatically projected at high velocity onto a surface. Shotcrete is also called spray-concrete, or spritzbeton (German).

Drainage

The presence of water within a rocky hillside is one of the major factors leading to instability. Knowledge of the water pressure and of the runoff mode is the basis for being able both to carry out credible stability analysis, and to plan measures aimed at improving hillside stability. Hoek and Bray (1981) provide a scheme of possible measures that can be actuated on a hillside to reduce not only the amount of water, which they believe in itself to be negligible as an instability factor, but above all the pressure brought to bear by the water. The proposed scheme was elaborated taking three principles into account:

- Preventing water entering the hillside through open or discontinuity traction cracks
- Reducing water pressure in the vicinity of potential breakage surfaces through selective shallow and sub-shallow drainage.
- Placing drainage in order to reduce water pressure in the immediate vicinity of the hillside.

The measures that can be achieved to reduce the effects of water can be shallow or in depth. Shallow drainage work has the main function of intercepting surface runoff water and keeping it away from potentially unstable areas. In reality, on rocky hillsides this type of measure although contributing to reducing the amount of infiltration, alone is insufficient to stabilise a hillside. Deep drainage is the most effective with this type of slope. Sub horizontal drainage is very effective in reducing pore-pressure along crack surfaces or potential breakage surfaces. In rocks the choice of drain spacing, slope, and length is subordinated to, apart from the hillside geometry, above all the structural formation of the mass. Features such as position, spacing and discontinuity opening persistence condition, apart from the mechanical characteristics of the rock, the water runoff mode inside the mass. Therefore, only by intercepting the mostly drained discontinuities can there be an efficient result. The sub horizontal drains are accompanied by surficial collectors which gather the water and take it away through networks of small surface channels.

Vertical drainage is generally associated with sunken pumps which have the task of draining the water and lowering the groundwater level. The use of continuous cycle pumps implies very high running costs conditioning the use of this technique for only limited periods. Drainage galleries are rather different in terms of efficiency. They are considered to be the most efficient drainage system for rocks even if they have the drawback of requiring very high technological and financial investment.

In particular, used in rocks this technique can be highly efficient in lowering water pressure. Drainage galleries can be associated with a series of radial drains which augment their efficiency. The positioning of this type of work is certainly connected to the local morphological, geological and structural conditions.

Geometry Modification

This type of measure is used in those cases in which, below the material to be removed, the rock face is sound and stable (for example unstable material at the top of the hillside, rock blocks thrusting out from the hillside profile, vegetation that can widen the rock joints, rock blocks isolated from the joints).

Detachment measures are carried out where there are risk conditions due to infrastructures or the passage of people at the foot of the hillside. Generally this type of measure can solve the problem by eliminating the hazard. However, it should be ensured that once the measure is carried out, the problem does not re-emerge in the short term. In fact, where there are very cracked rocks, the shallower rock portions can undergo mechanical incoherence, sometimes encouraged by extremes of climate, causing the isolation of unstable blocks.

The measure can be effected in various ways, which range from demolition with pick axes to the use of explosives. In the case of high and/or not easily accessible faces it is necessary to turn to specialists who work acrobatically.

When explosives are used, sometimes controlled demolition is needed, with the aim of minimising or nullifying the undesired effects resulting from the explosion of the charges, safeguarding the integrity of the surrounding rock.

Controlled demolition is based on the drilling of holes placed at a short distance from each other and parallel to the scarp to be demolished. The diametre of the holes generally varies from 40 to 80 mm; the spacing of the holes is generally about 10 to 12 times the diametre. The charge fuse times are established so that those at the outer edges explode first and the more internal ones successively, so that the area of the operation is delimited.

Protection Measures

The protection of natural and quarry faces can have two different aims:

- Protecting the rock from alterations,
- Protecting the infrastructure and towns from rockfalls.

It is therefore necessary to identify above all the cause of the alteration or the possibility of rockfall. Successively the area of operations needs to be delimited, the most suitable procedure to solve the problem and, finally, to control the effectiveness of the measure itself over a period of time. The most-used passive protection measures are:

- boulder-gathering trenches at the foot of the hillside,
- metal containment nets,
- boulder barriers.

As far as the boulder barriers are concerned, they are generally composed of suitably rigid metal nets. Moreover, lately, various structural types have been put on the market, for which the manufactures specify the kinetic energy of absorption. One of the structural control methods for boulder containment nets starts from the concept of projectile collision, on the basis of which the maximum applied force and the corresponding resultant buckling are expressed, by means of a static analysis, leading to the quantification of the forces divided up among the various structural elements. Another type of boulder containment barrier is the earth embankment, possibly reinforced with geo-synthetics (reinforced ground). The advantage of this type of work, compared to nets, is easier maintenance and lower environmental impact minor. The absorption of kinetic energy is generally greater than that of metal nets.

Soil Slopes

Geometry Modification: The operation of re-profiling a slope with the aim of improving its stability, can be achieved through various procedures:

- Lowering the slope
- Positioning infill at the foot of the slope

Slopes can be reduced by digging out the brow of the slope. This is effective for correcting shallow forms of instability, where movement is limited to layers of ground near to the surface and when the slopes are higher than 5m. Moreover, the steps created in this way and suitably achieved also reduce surface erosion.

However, caution should be exercised to avoid the onset of local breakage following the cuts made. Infill at the foot of the slope, instead, has a stabilising effect in the case of translational or deep rotational landslide, in which the landslide surface at the top submerges and describes a sub-vertical surface that re-emerges in the area at the foot of the slope. The choice of reducing the slope and infilling at the foot is rarely a problem since there are generally specific constraints to be respected at the top or at the foot of the slope.

Generally in slope stabilisation where there are no constraints (often this occurs for natural slopes) a combination of slope reduction and infilling at the foot of the slope is adopted to avoid heavy work of just one type. Included among work at the foot of the slope are the berm and some gravitational structures like gabions or reinforced ground, that is, concrete blocks. In the case of natural slopes the choice of re-profiling scheme is not so clear as in the case of artificial slopes. Often the profile is highly irregular with large areas of not recent natural creep, so that its shallow development can make some areas unserviceable as a cutting or infill point.

Where the buried shape of the old landslide is complicated, depositing of infill material in one area can lead to destabilising another.

When planning this type of work the stepping effect of the cuts and infill should be taken into account: their beneficial influence on the increase in Safety Factor will be reduced in relationship to the size of the landslide under examination. Moreover, it is very important to ensure that neither the cuts nor the infill mobilise the existing or potential creep plane of the landslide. Generally, infilling at the foot of the landslide should be preferred to cutting at the top (to reduce weight at the top of the slope), since the latter solution proves to be often more expensive than the former. Moreover, in complex and compound landslides, infill at the foot of the slope, at the tip of the foot itself, has a lesser probability of interfering with the interaction of the individual landslide elements.

An important aspect of stabilisation work that changes the morphology of the slope is that, in more precisely mechanical terms, effecting cuts and infill generates non-drained charge and discharge stresses. In fact, in the case of positioning infill, the safety factor SF, will be less in the short term than in the long term. In the case of effecting a cut in the slope, SF will be less in the long term rather than in the short term. Therefore in both cases it is opportune to calculate the SF both in the short and the long term.

Finally, it should be remembered that the effectiveness of infill increases with time on condition that it is associated with an appropriate infill drainage system, obtained through an underlying drainage cover or appropriate shallow drainage. More generally therefore re-profiling systems are associated with and integrated by surficial protection of the slope against erosion and by regulation of meteoric waters through drainage systems made up of ditches and small channels (clad or unclad and prefabricated) to run off the water collected.

These surficial water regulation systems are obtained by modelling the land itself around the body of the landslide large ground ditches in the case of incoherent material landslides) or by means of flexible suitably placed drainage pipes able to collect the water.

Figure: *Micro-perforated flexible drainage tubes*

These provisions will serve the purpose of avoiding penetration of the landslide body by circulating water or into any cracks or fissures, further decreasing ground shear strength. A problem that could be caused by water near the surface of the hillside is the erosion of surface material due to water runoff. This proves hazardous in terms of stability since it tends to weaken the slope by removing material as well as triggering excess pressure due to the water flow. For defence against erosion, a series of solutions can be used, such as:

- Geomats
- Geogrids
- Brushwood mats

These measures share the superficial character of their installation given their low environmental impact. Geomats or rather anti-eroding biomats or bionets are purpose-made synthetic products for the protection and grassing of slopes subject to surface wash through two main erosion control mechanisms: the containment and reinforcement of the surficial ground; the protection from the impact of the raindrops.

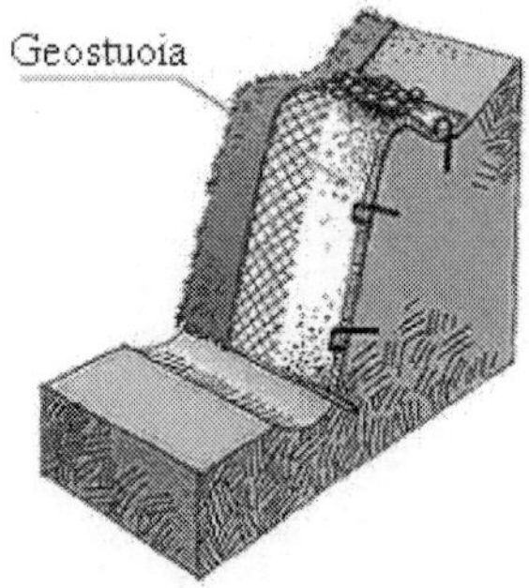

Figure: *typical anti-erosion solutions with geomats*

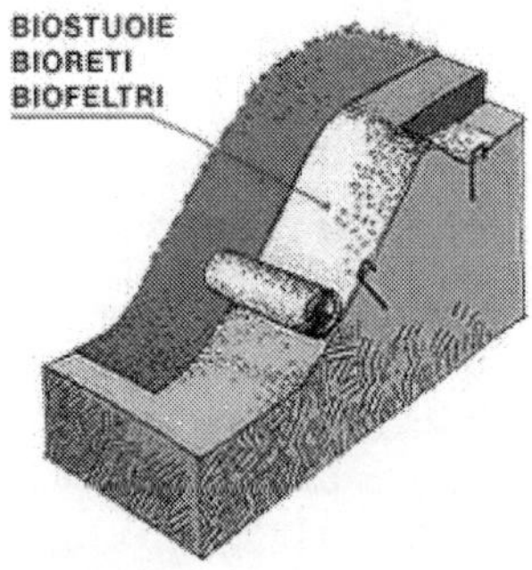

Figure: *Anti-erosion solutions with the use of bionets: typical scheme*

Figure: *Anti-erosion solutions with the use of bionets: types of bionets*

Geogrids - Wicker or brushwood mats are made of vegetal material. Very long and flexible willow branches can be used, which are then covered with infill soil. Alternating stakes of different woody species are used and they are woven to form a barrier against the downward drag of the material eroded by the free water on the surface

Figure: *typical brushwood solutions*

Draining Techniques

Drainage systems are adopted to reduce the neutral stresses in a potentially unstable hillside. In terms of safety for global stability, these measures translate into the lowering of the water level inside the mass, which consequently leads to reduction in pore-pressure in the ground and an increase in the shear strength available in particular along the potential creep surface. In relation to hillside morphology, the kinematics of movement predicted and to the depth of the creep surfaces, the reduction in pore-pressure by drainage can be obtained using shallow and deep drains. Usually, shallow drainage is adopted when the potential hillside movement is foreseen as shallow landslide affecting the ground to a depth of about 5-6m. When there is deeper surface slipping, deep drainage has to be introduced, but shallow drainage systems can be provided anyway with the aim of running off that aliquot of surficial water directly connected to seasonal rainfall.

Shallow Drainage

There are two types of shallow drainage solution:

- *Shallow drainage trenches:* Traditional drainage trenches are cut in an unbroken length and filled with highly permeable granular draining material;

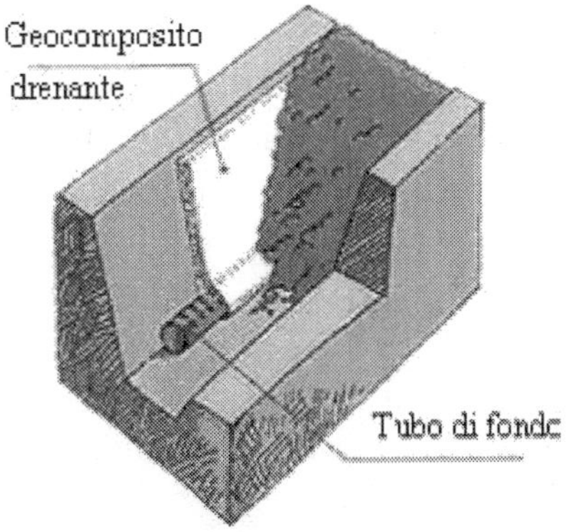

Figure: *Shallow drainage trenches equipped with geocomposites: typical scheme*

- *Shallow drainage trenches equipped with geocomposites:* These systems consist of unbroken trenches with scarped sides covered with geocomposites, generally with 25m. Long panels, having draining characteristics. The bottom of the trench houses a drainage tube with the task of bottom discharge placed in continuity to the geocomposite canvas.

Deep Drainage

Deep drainage acts by modifying the filtration routes in the ground. Often they are more expensive than shallow drains, but they are usually more effective because they remove the quantity of water that induces instability in the hillside, from within the ground and diminish the neutral stresses directly where necessary. Deep drainage in earth slopes can be obtained by means of the following works: large diametre drainage wells equipped with sub-horizontal drains. These systems can have just a structural function, just a draining function or both. The draining elements are the microdrains, perforated and positioned sub-horizontally and fanned out, oriented uphill to favour water discharge by gravity.

The size of the wells is chosen with the aim of allowing the insertion and functioning of the perforation equipment for the microdrains. Generally, the minimum internal diametre to be adopted must be greater than 3.5 m, for drains with a length of 20 to 30 m.

Longer drains require wells with a diametre of up to 8–10 m. To determine the network of microdrains planners have to take into consideration the makeup of the subsoil and the hydraulic regime of the slope, to provide for the correct number and distribution of the microdrains.

The drainage in these wells is passive, realised by linking the bottom of adjacent wells by sub-horizontal perforations (provided with temporary sheathing pipes) in which the microdrains are placed at a gradient of about 15-20° and are equipped with microperforated PVC pipes, protected by non-filtering fabric along the draining length.

Once all the drain is inserted in the hole and having embedded the latter in the ground, the temporary sheathing is completely removed and the head of the drain is cemented to the well. In this way a discharge line is created linking all the wells emerging to the surface downhill, where the water is discharged naturally without the help of raising pumps.

The wells are placed at such a distance apart that the individual collecting areas of the microdrains, appertaining to each well, are overlaid. In this way all the volume of the slope involved with the water table is drained. Medium-diametre drainage wells linked at the bottom. The technique involves the dry cutting with temporary sheathing pipes, of aligned drainage wells, with a diametre of 1200–1500 mm., positioned at an interaxis of 6–8 m., their bottoms linked together to a bottom tube for the discharge of drained water. In this way the water discharge takes place passively, due to gravity by perforated pipes with mini-tubes, positioned at the bottom of the wells themselves. The linking pipes, generally made of steel, are blind in the linking length and perforated or windowed in the length corresponding to the well. The wells have a concrete bung at the bottom and are filled, after withdrawal of the temporary sheathing pipe, with dry draining material and are closed with an impermeable clay bung.

In normal conditions, these wells reach a depth of 20–30 m, but, in especially favourable cases, a depth of even 50 m can be reached. Some of these wells have drainage functions across their whole section and others can be inspected. The latter serve for maintenance of the whole drainage screen. Such wells that can be inspected are also a support point for the creation of new drainage wells and access for the installation, also on a later occasion, for a range of sub-horizontal drains at the bottom or along the walls of the wells themselves, with the purpose of increasing the drainage capacity of the well.

Isolated Wells Fitted with Drainage Pumps

This system provides for the installation of a drainage pump for each well. The distribution of the wells is established according to the permeability of the land to be drained, the lowering of the water pressure to be achieved and the area that has to be involved in this work, so that the water pressure area deriving from the depression fans generated by the single drainage points, responds to the needs of the plan. The use of isolated wells with a drainage pumps leads to high running costs and imposes a very time-consuming level of control and maintenance.

Deep Drainage Trenches

Deep drainage trenches consist of unbroken cuts with a small cross-section that can be covered at the bottom with geofabric canvas having a primary filter function. They are filled with draining material that has a filtering function and exploits the passive drainage to carry away the drained water downhill. The effectiveness of these systems is connected to the geometry of the trench and the continuity of the draining material along the whole trench. As far as the geometry of the cut is concerned attention should be paid to the slope given to the bottom of the cut. In fact, deep drainage trenches do not have bottom piping that is inserted in the end part of the trench, downhill, where the depth of the cut is reduced until the campaign level is reached.

Drainage Galleries Fitted with Microdrains

Drainage galleries constitute a rather expensive stabilisation provision for large, deep landslide movements, to be carried out where the ground is unsuitable for cutting trenches or drainage wells and when it is impossible to work on the surface owing to a lack of space for the work machinery. Their effectiveness is due to the extensiveness of the area to be drained. Moreover, these drainage systems have to be realised on the stable part of the slope.

The drainage systems are placed inside the galleries and are made up of microdrains, with lengths that can reach 50–60 m and are spatially oriented in a suitable directions. The sizes of the galleries are conditioned by the need to insert the drain perforation equipment. For this reason the minimum transversal internal size of the galleries vary from a minimum of 2 m, when using special reduced size equipment, and to at least 3.5 m, when using traditional equipment.

Siphon Drain

This is a technique conceived and developed in France, which works like the system of isolated drainage wells but overcoming the inconvenience of installing a pump for each well. In fact, the system on the principle that once motion is triggered in the siphon tube, avoiding the entry of air into the loop, the motion is uninterrupted.

For this reason, in this system, the condition that the two ends of the siphon tube are submerged in the water of the two permanent storage tanks, must always be checked. This drain is created vertically starting from the campaign level but can also be sub-vertical or inclined. The diametre of the well can vary from 100 to 300 mm; inside a PVC pipe is placed or a perforated or microperforated steel pipe, filled with draining material. The siphon drain in this way carries off of drainage water by gravity without the need for drainage pumps or pipes linking the bottom of each well. This system proves to be economically advantageous and relatively simple to set up even if it necessitates a programme of controls and maintenance.

Microdrains

Microdrains is a simple to create drainage system with contained costs. They consist of small diametre perforations, made from surface locations, in trenches, in wells or in galleries. The microdrains are set to work in a sub-horizontal or sub-vertical position, according to the type of application.

Reinforcement Measure

The stabilisation of the hillside in terms of an increase in the mechanical characteristics of the potentially unstable ground, can be effected by means of two different approaches:

- Insertion of reinforcement elements in the ground;
- The improvement of the mechanical characteristics of the ground volume affected by landslide through chemical, thermal or mechanical treatment.

Insertion of Reinforcement Elements in the Ground

This category of work uses:

- Large diametre wells supported by one or more crowns of consolidated and possibly reinforced earth columns;
- Anchors
- Networks of micropiles
- Soil nailing
- Geogrids for reinforced ground
- Cellular faces

Large diametre wells supported by one or more crowns of consolidated and possibly reinforced earth columns To guarantee slope stability it may be necessary to insert very rigid, strong element.

These elements are large diametre full section or ring section reinforced concrete wells. The wells can have a circular or elliptical section. The depth of the static wells can reach 30-40m. Often the static stabilising action of the wells is integrated with a series of microdrains laid out radially on several levels, also to reduce pore-pressure, if it is hazardous.

Earth Anchor

An earth anchor is a device that is designed to support structures and is used in geotechnical and construction applications. It can also be referred to as a ground anchor, percussion driven earth anchor or mechanical anchor.

Earth anchors can be used in either temporary or permanent applications. Typical uses for earth anchors include for supporting retaining walls and for crop protection structures.

Typical Applications

Figure: *An earth anchor for guy wires on a guyed mast near Thabazimbi, South Africa*

Typically used within civil engineering and construction projects, earth anchors are used in a variety of applications:

- Earth anchors have had much use in the retaining of slopes. Typically, earth anchors have been used in this context to prevent deep seated failures, as well as a form of erosion control.
- Another typical usage of Earth Anchors is the structural support of Temporary Buildings and Structures. Such structures include circus tents and outdoor stages.

- Although unrelated to the typical boat anchor, earth anchors have also been used within marine applications. These applications include structural support for floating docks and pipelines.
- Another usage of ground anchors has been in the support of guyed structures.
- Earth anchors have also been used for general security, for example, in the anchoring of small aircraft.
- One more commonly seen usage of mechanical earth anchors is within utility systems. Earth anchors are typically used for the anchoring of utility poles.
- Some anchor systems have now been combined with drainage solutions. This allows for loadlocking and restraining capability to happen simultaneously.
- Earth anchors have also had great success in the anchoring of trees. This anchoring is typically used when a semi-mature tree is transplanted to a different site. Earth anchors have popularity for being used in this area as anchoring trees by using this method removes the need for unsightly guy wires and tree stakes.

Installation

Before installing an earth anchor, site analysis has to be done at the site of installation. This analysis is done so that it can be determined what loading capacity is required of the earth anchors. This site analysis needs to include the depth that the earth anchor is driven into, the soil strength and the moisture content of the soil. The site analysis also needs to measure the corrosivity of the soil, so that a suitable anchor material can be picked which matches the design life of the project.

Earth anchors normally have a drive rod attached. The earth anchor is driven into the ground, typically with an impact hammer. However, this installation method may differ depending on the composition of the soil that the anchor is being inserted into. In more dense soils, a pilot hole is drilled into the earth. Typically, anchors are inserted into the soil and used for testing before the final anchor design is decided on. After an anchor has been installed, the drive rod is removed and the anchor is load-locked. For lighter anchors, this is typically done by a hand tool. The angle of the wings on the anchor allow for increase of frustum pressure on the soil. Normally, earth anchors are designed to rotate ninety degrees during this load-locking process.

How Earth Anchors Work

Once load-locked, the earth anchor exerts effort to the soil above the anchor. This pressure is in a conical shape, with the size and angle of this conical shape depending on factors including:

- The shear angle of the soil.
- The depth at which the earth anchor has been installed.
- The load applied to the earth anchor.
- The size of the earth anchor.

This conical compression of the soil creates resistance to any further movement of the anchor. The shape of Earth Anchors are typically designed to provide as much frustum shaped pressure as possible, and this is typically achieved with the usage of large surface areas and trailing edges around the anchor.

Networks of Micropiles

This solution provides for a plant of a series of micropiles that make up a three-dimensional grid, variably tilted and linked at the head by a rigid reinforced concrete mortise. This structure constitutes a reinforcement for the ground, inducing an intrinsic improvement of the ground characteristics incorporated in the micropiles. A measure of this type proves effective in cases of not very extensive landslide.

The effectiveness of solutions with micropiles is linked to the possibility of inserting micropiles over the whole width of the landslide area. In the case of rotational landslides in soft clay, the piles contribute to increasing the resisting moment by friction on the upper part of the pile shaft found in the landslide.

This functionality is usually valid for micropiles working using the point. In the case of suspended piles, strength is governed by the part of the pile offering the least resistance. As an operational method, those piles that destined to the most unstable area of the slope should be positioned first, subsequently those around the unstable area, in order to reduce any possible lateral ground displacements.

Preliminary design methods for the micropiles, nowadays are entrusted to the use of codes that carry out numerical simulations, but which are subject to simplifications in the models that necessitate characterisations of rather precise potential landslide material.

Soil Nailing

Soil nailing is a construction technique that can be used as a remedial measure to treat unstable natural soil slopes or as a construction technique that allows the safe over-steepening of new or existing soil slopes. The technique involves the insertion of relatively slender reinforcing elements into the slope – often general purpose reinforcing bars (rebar) although proprietary solid or hollow-system bars are also available. Solid bars are usually installed into pre-drilled holes and then grouted into place using a separate grout line, whereas hollow bars may be drilled and grouted simultaneously by the use of a sacrificial drill bit and by pumping grout down the hollow bar as drilling progresses. Kinetic methods of firing relatively short bars into soil slopes have also been developed. Bars installed using drilling techniques are usually fully grouted and installed at a slight downward inclination with bars installed at regularly spaced points across the slope face. A rigid facing (often pneumatically applied concrete, otherwise known as shotcrete) or isolated soil nail head plates may be used at the surface. Alternatively a flexible reinforcing mesh may be held against the soil face beneath the head plates. Rabbit proof wire mesh and environmental erosion control fabrics and may be used in conjunction with flexible mesh facing where environmental conditions dictate.

Soil nail components may also be used to stabilize retaining walls or existing fill slopes (embankments); this is normally undertaken as a remedial measure.

Since its first application using modern techniques in Versailles in 1972, soil nailing is now a well-established technique around the world. One of the first national guideline publications for soil nailing was produced in Japan in 1987. The U.S. Federal Highway Administration issued guideline publications in 1996 and 2003.

Preliminary Analysis

Four main points to be considered in determining if soil nailing would be an effective retention technique are as follows. First, the existing ground conditions should be examined. Next, the advantages and disadvantages for a soil nail wall should be assessed for the particular application being considered. Then other systems should be considered for the particular application. Finally, cost of the soil nail wall should be considered. Soil nail walls can be used for a variety of soil types and conditions. The most favourable conditions for soil nailing are as follows: The soil should be able to stand unsupported one to two metres high for a minimum of two days when cut vertical or nearly vertical. Also all soil nails within a cross section should be located above the groundwater table. If the soil nails are not located above the groundwater table, the groundwater should not negatively affect the face of the excavation, the bond between the ground and the soil nail itself. Based upon these favourable conditions for soil nailing stiff to hard fine-grained soils which include stiff to hard clays, clayey silts, silty clays, sandy clays, and sandy silts are preferred soils. Sand and gravels which are dense to very dense soils with some apparent cohesion also work well for soil nailing. Weathered rock is also acceptable as long as the rock is weathered evenly throughout (meaning no weakness planes). Finally, glacial soils work well for soil nailing.

A list of unfavourable or difficult soil conditions for soil nailing can include dry, poorly graded cohesion-less soils, soils with a high groundwater table, soils with cobbles and boulders, soft to very soft fine-grained soils, highly corrosive soils, weathered rock with unfavourable weakness planes, and loess. Other difficult conditions include prolonged exposure to freezing temperatures, a climate that has a repeated freeze-and-thaw cycle, and granular soils that are very loose.

Origins

Soil nailing evolved from the New Austrian Tunneling method which is a system for underground excavations in rock. This method consists of passive steel reinforcement in the rock followed by the application of reinforced shotcrete. This concept of combining passive steel reinforcement and shotcrete has also been applied to the stabilization of rock slopes since the early 1960s.

The first application of soil nailing was implemented in 1972 for a railroad widening project near Versailles, France. Soil nails were used to stabilize an 18 m high slope consisting of sandy soil. This method proved to be more cost-effective, while at the same time cut down the construction time when compared to other conventional support methods. Germany was the next country to investigate soil nailing. From 1975-1981 the University of Karlsruhe and the construction company Bauer collaborated to establish a research programme. This programme conducted full scale testing of experimental walls with different configurations and developed analysis procedures for use in design. The United States first used soil nailing in 1976 for the support of a 13.7 m deep foundation excavation in dense silty sands. Soil nailing was implemented in the expansion of The Good Samaritan Hospital in Portland, Oregon. This retaining system was produced in approximately half the time at about 85% of the cost of conventional retaining systems.

Design

After a preliminary analysis of the site, initial designs of the soil nail wall can be begin. This begins with a selection of limit states and design approaches. The two most common limit states used in soil nail wall design is strength limit and service limit states. The strength limit state is the limit state that addresses potential failure mechanisms or collapse states of the soil nail wall system. The service limit state is the limit state that addresses loss of service function resulting from excessive wall deformation and is defined by restrictions in stress, deformation and facing crack width under regular service conditions. The two most common design approaches for soil nail walls are limit state design and service load design.

Initial design considerations include wall layout (wall height and length), soil nail vertical and horizontal spacing, soil nail pattern on wall face, soil nail inclination, soil nail length and distribution, soil nail material and relevant ground properties. With all these variables in the mind of the design engineer the next step is to use simplified charts to preliminarily evaluate nail length and maximum nail force. Nail length, diametre and spacing typically control external and internal stability of the wall. These parametres can be adjusted during design until all external and internal stability requirements are met. After the initial design is completed, final design progresses where the soil nail wall has to be tested for external and internal failure modes, seismic considerations and aesthetic qualities. Drainage, frost penetration and external loads such as wind and hydrostatic forces also have to be determined and included in the final examination of the design.

Construction

With the design complete, construction is the next step. Most soil nail wall construction follows a specific procedure. First a cut is excavated and temporary bracing is put in place if necessary. This is done with conventional earth moving equipment and hydraulic drills. Next, holes for the soil nails are drilled at predetermined locations as specified by the design engineer. The equipment used for this step is dependent on the stability of the material in which the soil nail wall is supporting. Rotary or rotary percussive methods using air flush or dry auger methods can be used with stable ground. For unstable ground, single tube and duplex rotary methods with air and water flush or hollow stem auger methods are used. With the holes drilled, the next step is to install and grout the nails into place. After all nails are inserted, a drainage system is put into place. Synthetic drainage mat is placed vertically between the nail heads, which are extended down to the base of the wall where they are most commonly connected to a footing drain. A layer of shotcrete is applied and bearing plates are installed before a final facing is put in place to complete the soil nail wall. Variations of the steps described above may be necessary to accommodate additional preparation tasks or supplementary activities for specific project conditions.

In terms of construction, soil nail walls have a decisive advantage over other alternatives. Soil nail walls require a smaller right-of-way than ground anchor walls and have less environmental impact. Installation of soil nail walls is relatively rapid and typically uses less materials and smaller construction equipment than ground anchor walls.

Cost Comparison

One great advantage of soil nail walls is their cost-effectiveness over other alternatives. When conventional soil nailing construction procedures are used, soil nail walls are much more economical than concrete gravity walls and similarly or more cost effective than ground anchor walls.

Inspection and Performance Monitoring

Inspection activities play a vital role in the production of high-quality soil nail walls because conformance to project plans and specifications should result in a soil nail wall that will perform its intended duty for its designed duration. Inspections usually involve evaluation of the following: conformance of system components to material specification, conformance of construction methods to execution specifications, conformance to short-term performance specifications, and long-term monitoring. Short-term performance specifications are checked with loads tests, which utilize hydraulic jacks and pumps to perform several load applications. Three common load tests for short-term performance are verification or ultimate load tests, proof tests and creep tests. Verification or ultimate load tests are conducted to verify the compliance of the soil nails with pullout capacity and strengths resulting from the contractor's installation method. Proof tests are intended to verify that the contractor's construction procedure has been consistent and that the nails have not been drilled and grouted in a soil zone not tested in the verification stage. Creep tests are performed to ensure that the nail design loads can be safely carried throughout the structure's service life.

Long term performance monitoring is used to collect data to ensure adequate performance and refine future design practices. Parametres to be measured include vertical and horizontal movement of the wall face, local movements or deterioration of facing elements, drainage to the ground, loads, load distribution and load changes in the nails, temperature and rainfall. These parametres are measured using several specific tools including inclinometres, load cells and strain gauges.

Geogrids

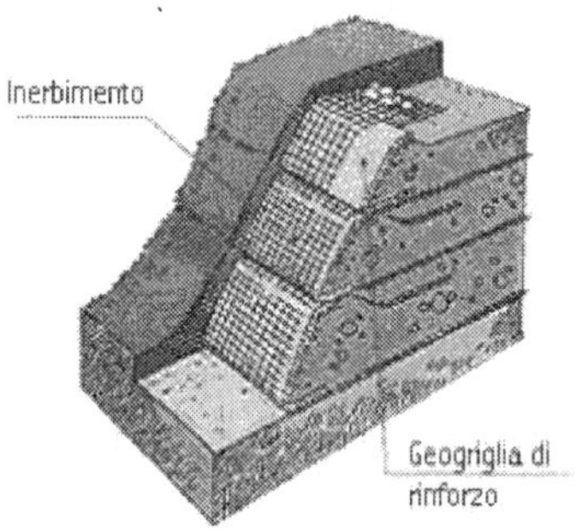

Figure: *typicalgeogrid solution*

The geogrids are synthetic materials used to reinforce the ground.

The insertion of geosynthetic reinforcements (generally in the direction in which the deformation has developed) therefore has the function of reinforcing the ground conferring greater stiffness and stability upon it and the capacity to be subjected to greater deformations without reaching fracture point.

Cellular Faces

Cellular faces, also known by the name of "crib faces" are special supporting walls realised by means of head grids prefabricated in reinforced concrete or in wood (treated externally with preservatives). The heads have a length of about 1–2 m and the work can also reach 5 m in height.

Compacted granular material is inserted in the spaces of the grid. The characteristic modularity of the system confers notable flexibility of use on its use both in terms of adaptability to the morphology of the ground and because the structure does not require any deep foundation other than a laying plane of lean concrete useful to make the support plane of the whole structure regular. This solution can also take vegetal ground into the grid spaces, to favour the camouflaging of the work into the surroundings by means of the vegetation rooting on the exposed face.

Improvement in the Mechanical Characteristics of the Soil Volume Affected by Landslide by Means of Chemical, Thermal or Mechanical Treatment

Among the treatments belonging to the group of intrinsic improvements of a mechanical character in the ground, the technique of jet-grouting is often used, in particular as a substituting and/or integrating element for technologies used for the previously described structural measures. The phases of this work are:

Phase I

Perforation phase: insertion, with perforation destroying the nucleus, of a set of poles into the ground up to the depth of treatment required by the project.

Phase II

Extraction and programmed injection phase: injection of the mixture at very high pressure is done during the extraction phase of the set of poles. It is in this phase that through the insistence of the jet in a certain direction for a certain interval of time, the effect is obtained by the speed of extraction and rotation of the set of poles, so that volumes of ground can be treated in the shape and size desired.

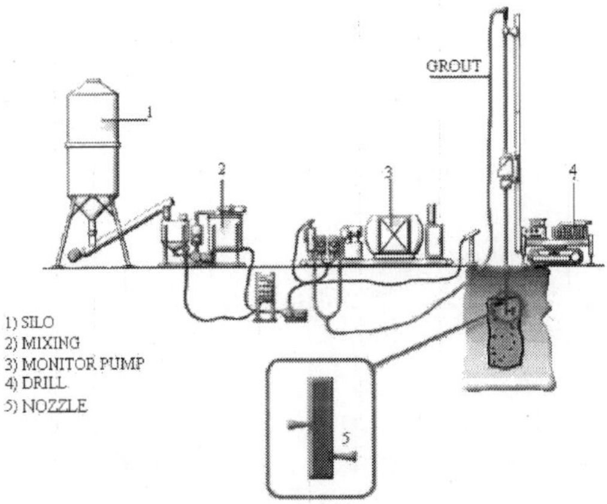

Figure: *jet-grouting device*

The high energy jet produces a mixture of the ground and a continuous and systematic "claquage" with only a local effect within the radius of action without provoking deformations at the surface that could induce negative consequences on the stability of adjacent constructions. The projection of the mixture at high speed through the nozzles, using the effect of the elevated energy in play, allows the modification of the natural disposition and mechanical characteristics of the ground in the desired direction and in accordance with the mixture used (cement, bentonite, water, chemical, mixtures etc.). Depending on the characteristics of the natural ground, the type of mixture used, and work parametres, compression strength from 1 to 500 kgf/cm^2 (100 kPa to 50 MPa) can be obtained in the treated area.

The realisation of massive consolidated ground elements of various shapes and sizes (buttresses and spurs) within the mass to be stabilised, is achieved by acting opportunely on the injection parametres. In this way the following can be obtained: thin diaphragms, horizontal and vertical cylinders of various diametre and generally any geometrical shapes.

Another improvement system of the mechanical characteristics of the ground is that of thermal treatment of potentially unstable hillsides made up of clayey materials. Once on the railways, for the treatment of unstable slopes in clayey grounds, the lighting of wood or coal fires was used in holes dug into the slope, to harden the surrounding ground. In large diametre holes (from 200 to 400 mm.), about 0.8-1.2m. apart and horizontally interconnected, burners were introduced to form cylinders of hardened clay. The temperatures reached were around 800°C. These clay cylinders worked like piles giving greater shear strength to the creep surface. This system was useful for surface creep, as in the case of an embankment. In other cases the depth of the holes or the amount of fuel necessary led to either the exclusion of this technique or made the effort ineffective.

Other stabilisation attempts were made by using electro-osmotic treatment of the ground. This type of treatment is applicable only in clayey grounds. It consists of subjecting the material to the action of a continuous electrical field, introducing pairs of electrodes embedded in the ground. These electrodes, when current is introduced cause the migration of the ion charges in the clay. Therefore the inter-pore waters are collected in the cathode areas and they are dragged by the ion charges. In this way a reduction in water content is achieved. Moreover, by suitable choice of anodic electrode a structural transformation of the clay can be induced due to the ions freed by the anode triggering a series of chemo-physical reactions improving the mechanical characteristics of the unstable ground.

This stabilisation method, however, is effective only in homogeneous clayey grounds. This condition is hard to find in unstable slopes, therefore electro-osmotic treatment, after some applications, has been abandoned.

Soil Slope Stabilization

Coir Geotextiles are another product which is used for erosion control globally and is used for bioengineering and slope stabilization applications due to the mechanical strength necessary to hold soil together. It is been proven that coir geotextiles last for approximately 3 – 5 years depending on the weight, and by the time the product degrades, it converts itself it to humus, which enriches the soil.

With Steel Wire Mesh

Steel wire mesh may be used for soil and rock slope stabilization. After leveling, the surface is covered by a steel-wire mesh, which is fastened to the slope and tensioned. It is a cost-effective approach.

Cellular Confinement

Cellular Confinement Systems (CCS, also known as geocells) are widely used in construction for erosion control, soil stabilization on flat ground and steep slopes, channel protection, and structural reinforcement for load support and earth retention. Typical cellular confinement systems are made with ultrasonically-welded high-density polyethylene (HDPE) or Novel Polymeric Alloy strips that are expanded on-site to form a honeycomb-like structure which may be filled with sand, soil, rock or concrete.

History of Cellular Confinement

Research and development of cellular confinement systems (CCS) began with the U.S. Army Corps of Engineers in September 1975 to test the feasibility of constructing tactical bridge approach roads over soft ground. Engineers discovered that sand-confinement systems performed better than conventional crushed stone sections. They concluded that a sand-confinement system could be developed that would provide an expedient construction technique for building approach roads over soft ground and that the system would not be adversely affected by wet weather conditions. These early efforts led to the civilian commercialization of the first cellular confinement system known as Geoweb® by the Presto Products Company. The cellular confinement system was made from high density polyethylene (HDPE) that was light weight, strong and durable. This new Geoweb cellular confinement system was used first for load support applications in the United States in the early 1980s; second for slope erosion control and channel lining in the United States in 1984 and; third for earth retention in Canada in 1986. Research on cellular confinement in these application areas in cooperation with Presto Products also started during the 1980s.

Research by Drs. Bathurst and Jarrett discovered that cellular confinement reinforced gravel bases are "equivalent to about twice the thickness of unreinforced gravel bases" when placed over a saturated peat sub-base. Further, 1.25 mm (50 mil) HDPE performed better than single sheet reinforcement schemes (geotextiles and geogrids) and was more effective in reducing lateral spreading of the infill material under loading than conventional reinforced bases. In terms of the effectiveness of confinement, geocells have more attractive features due to its 3D structure than any other planar geosynthetic reinforcement. Since this early work, the results of large-scale triaxial test on isolated geocells has demonstrated that cellular confinement imparts apparent cohesion to cohesionless compacted granular material on the order of 169 kPa - 190 kPa (3500 psf - 4000 psf). Cellular confinement systems are now recognized as an important technology when applied to load support (Webster, 1986 and Bathurst & Jarrett, 1988) under roads and rail lines, gravity and reinforced earth retaining wall systems (Crowe, Bathurst & Alston, 1989), (Bathurst, Crowe &Zehaluk, 1993), slope stabilization and erosion control, channel lining systems (Engel, P. &Flato, G. 1987) (Simons, Li & Associates, 1988) (Wu & Austin, 1992) and other innovative uses. Cellular Confinement Systems (CCS, also known as geocells) are widely used in construction for erosion control, soil stabilization on flat ground and steep slopes, channel protection, and structural reinforcement for load support and earth retention. Typical cellular confinement systems are made with ultrasonically-welded high-density polyethylene (HDPE) or Novel Polymeric Alloy strips that are expanded on-site to form a honeycomb-like structure which may be filled with sand, soil, rock or concrete.

Recent Developments in Cellular Confinement Technology

Despite the effectiveness of the geocell technology, particularly in slope and channel applications, its use in base reinforcement of paved roads and railways was limited due to the lack of design methods, lack of advanced research in the last two decades and limited understanding of the reinforcement mechanisms (Yuu, et al.

2008).) Recent research in the last few years on geocell reinforcement for roadway applications - reflected by some 40 published papers - has been conducted at the University of Kansas as well as at other leading research institutes around the world, to understand the mechanisms and influencing factors of geocell reinforcement, evaluate its effectiveness in improving roadway performance, and develop design methods for roadway applications (Han, et al. 2011).

Research was conducted on HPDE geocells as well as geocells manufactured from a novel polymeric alloy (NPA), called Neoloy, developed by PRS. NPA is a composite polymeric alloy based on nano-fibres (polyester and nylon) in a polyolefin matrix. The NPA combines the desired properties of polyethylene and polyester, thus enabling a more effective use of geocells in new critical applications, such as reinforcement for earth retention, load support in pavements and railroads and more (Leshchinsky, et al., 2009). While HDPE is the commonly used material for geocells, leading researchers have questioned its suitability for long term applications (Leshchinsky, et al., 2009). This concern is backed up by "facts on the ground" as HDPE geocells are rarely used in critical applications, such as in the base layer of major highways and railways, subject to long-term heavy static and dynamic loading.

Laboratory plate loading tests on geocells showed that the performance of geocell-reinforced bases depends on the elastic modulus of the geocell. The geocell with a higher elastic modulus had a higher bearing capacity and stiffness of the reinforced base. Geocells made from NPA were found significantly better in ultimate bearing capacity, stiffness, and reinforcement relative to geocells made from HDPE (Pokharel, et al., 2009). NPA geocells showed better creep resistance and better retention of stiffness and creep resistance particularly at elevated temperatures, verified by plate load testing, numerical modeling and full scale trafficking tests (Pokharel, et al. 2011). Research demonstrated that NPA geocells have a lower thermal expansion coefficient and creep reduction factor, and higher tensile stiffness and strength than HDPE geocells. (Thakur, et al., 2010); and NPA increased the bearing capacity and reduced settlement of compacted sand base courses significantly more than geocells fabricated from HDPE (Pokharel, 2011, et al.).

Laboratory studies, full-scale moving wheel tests, and field demonstrations (cosponsored by US DOT Department of Transportation as well as state DOTs) have demonstrated clear benefits of NPA (novel polymeric alloy) geocell reinforcement in terms of increased stiffness and bearing capacity, wider stress distribution, reduced permanent deformation, and prolonged roadway life, while the design methods developed and calibrated in this research can help engineers design future roadway applications using geocells (Han, et al. 2011). This close cooperation and iterative research and development process between private industry and academia was cited by the editor of *Geosynthetics* magazine, as: "*an example of how product development for the geosynthetics industry can be done effectively... and can further advance the geosynthetics industry into the 21st century with much success.*"

Application vs. Long-Term Performance

HDPE-based geocells have been successfully installed in thousands of projects worldwide. However, it is incumbent to differentiate between low load applications, such as slope and channel applications, and new heavy-duty applications, such as in the base layer of asphalt pavement structures of heavily trafficked motorways and highways. While all polymeric materials used in geocells creep over time and under loading, the question is; what is the rate of degradation, under which conditions, how will this impact performance, and when will it fail?

The lifespan of geocells in slope protection applications, for example, is less critical as vegetative growth and root interlock stabilize the soil. This in effect compensates for any long-term loss of confinement in the geocells. Similarly, load support applications for low volume roads that are not subject to heavy loading usually have a short design life; therefore any minor loss of performance is tolerable. However, in critical applications such as reinforcement of the structural layer of asphalt highway pavements, long term dimensional stability is critical. The required design life for such roads under heavy traffic loads is typically 20–25 years, requiring verifiable long-term durability.

The Development of Standards for Testing Geocells

Standards for geocells have not kept pace with the developments in the field of testing for material sciences, ignoring ASTM and ISO methods for testing, verification and quality assurance of polymer plastics, such as TMA - Thermomechanical analysis, DMA - Dynamic Mechanical Analysis, Stepped Isothermal Method (SIM) and CTE - Coefficient of Thermal expansion. These methods are particularly suited for predicting long-term behaviour and accumulated plastic strain in a geosynthetic under loading under different mechanical stresses, frequencies and temperatures. These widely accepted testing methods are used by the pipe, automobile, electronic, military, security and construction industries. Geomembrane testing, for example, utilizes accelerated test methods, which use temperature to stimulate aging over time to evaluate their durability.

Unfortunately, these ASTM/ISO procedures commonly utilized by many other industries to evaluate performance have not been adopted by the most of geocell industry. Current standards evolved from the world of 2D planar geosynthetics. These do not fully reflect the composite behaviour of 3D geometry in soil, nor do they test long-term parametres such as: dynamic loading, permanent plastic deformation, effect of temperatures, environmental durability, etc. Therefore, new standards for geocells were proposed and under discussion by leading experts in geosynthetics in ASTM technical committee D-35. The goal is to set new industry standards that more accurately reflect 3D geocell geometry and material performance in the field rather than lab tests of individual strips and virgin materials that are used by most manufacturers today.

How it Works

A Cellular Confinement System when infilled with compacted soil creates a new composite entity that possesses enhanced mechanical and geotechnical properties. When the soil contained within a geocell is subjected to pressure, it causes lateral stresses on perimetre cell walls. The 3D zone of confinement reduces the lateral movement of soil particles while vertical loading on the contained infill results in high lateral stress and resistance on the cell-soil interface. These increase the shear strength of the confined soil, which:

- Creates a stiff mattress or slab to distribute the load over a wider area
- Reduces punching of soft soil
- Increases shear resistance and bearing capacity
- Decreases deformation

Confinement from adjacent cells provides additional resistance against the loaded cell through passive resistance, while lateral expansion of the infill is restricted by high hoop strength. Compaction is maintained by the confinement resulting in long term reinforcement.

Applications

Load Support: Cellular Confinement Systems (CCS) have been used to improve the performance of both paved and unpaved roads by reinforcing the soil in the subgrade-base interface or within the base course. The effective load distribution of CCS creates a strong, stiff cellular mattress. This 3D mattress reduces vertical differential settlement into soft subgrades, improves shear strength, and enhances load-bearing capacity, while reducing the amount of aggregate material required to extend the service life of roads. As a composite system, cellular confinement strengthens the aggregate infill, thereby simultaneously enabling the use of poorly graded inferior material (e.g. local native soils, quarry waste or recycled materials) for infill as well as reducing the structural support layer thickness. Typical load support applications include reinforcement of base and subbase layers in flexible pavements, including: asphalt pavements; unpaved access, service and haul roads; railway substructure and ballast confinement; working platforms in intermodal ports; airport runways and aprons, permeable pavements; pipeline road support; green parking facilities and emergency access areas.

Slope and Channel Protection

The three-dimensional lateral confinement of CCS along with anchoring techniques ensures the long-term stability of slopes using vegetated topsoil, aggregate or concrete surfacing (if exposed to severe mechanical and hydraulic pressures). The enhanced drainage, frictional forces and cell-soil-plant interaction of CCS prevents downslope movement and limits the impact of raindrops, channeling and hydraulic shear stresses. The perforations in the 3D cells allow the passage of water, nutrients and soil organisms. This encourages plant growth and root interlock, which further stabilizes the slope and soil mass, and facilitates landscape rehabilitation. Typical applications include: construction cut and fill slopes and stabilization; road and rail embankments; pipeline stabilization and storage facility berms; quarry and mine site restoration; channel and coastline structures.

Earth Retention

CCS systems provide steep vertical mechanically stabilized earth structures (either gravity or reinforced walls) for steep faces, walls and irregular topography. Construction of CCS earth retention is simiplified as each layer is structurally sound thereby providing access for equipment and workers, while eliminating the need for concrete formwork and curing. Local soil can be used for infill when suitable and granular, while the outer faces enable a green or tan fascia of the horizontal terraces/rows utilizing topsoil. Walls also can be used for lining channels and in cases of high flow, it is required that the outer cells contain concrete or cementious slurry infill.

Reservoirs and Landfills

CCS provides membrane liner protection, while creating stable soil, berms and slopes, for non-slip protection and durable impoundment of liquids and waste. Infill treatment depends on the contained materials: concrete for ponds and reservoirs; gravel for landfill drainage and leachates, vegetated infill for landscape rehabilitation. Concrete work is efficient and controlled as CCS functions as ready-made forms; CCS with concrete forms a flexible slab that accommodates minor subgrade movement and prevents cracking. In medium and low flow-velocities, CCS with geomembranes and gravel cover can be used to create impermeable channels, thereby eliminating the need for concrete.

Sustainable Construction

CCS is a green solution that makes civil infrastructure projects more sustainable. In load support applications, by reducing the amount and type of infill needed to reinforce soil, the usage of haul and earthmoving equipment is reduced. This in turn decreases fuel use, pollution and the carbon footprint, and at the same time minimizes on-site disruption from dust, erosion and runoff. When used for slope applications, perforated geocells provides excellent soil protection, water drainage and growth stratum for plants. The long-term design life of advanced CCS technology means that maintenance and the associated environmental costs are significantly reduced, as are long-term economic costs.

Avalanche

An avalanche (also called a snowslide or snowslip) is a rapid flow of snow down a slope. Avalanches are typically triggered in a starting zone from a mechanical failure in the snowpack (slab avalanche) when the forces on the snow exceed its strength but sometimes only with gradually widening (loose snow avalanche). After initiation, avalanches usually accelerate rapidly and grow in mass and volume as they entrain more snow. If the avalanche moves fast enough some of the snow may mix with the air forming a powder snow avalanche, which is a type of gravity current.

Slides of rocks or debris, behaving in a similar way to snow, are also referred to as avalanches. The remainder of this article refers to snow avalanches.

The load on the snowpack may be only due to gravity, in which case failure may result either from weakening in the snowpack or increased load due to precipitation. Avalanches that occur in this way are known as spontaneous avalanches. Avalanches can also be triggered by other loads such as skiers, snowmobilers, animals or explosives. Seismic activity may also trigger failure in the snowpack and avalanches.

Although primarily composed of flowing snow and air, large avalanches have the capability to entrain ice, rocks, trees, and other material on the slope, and are distinct from mudslides, rock slides, and serac collapses on an icefall. Avalanches are not rare or random events and are endemic to any mountain range that accumulates a standing snowpack. Avalanches are most common during winter or spring but glacier movements may cause ice and snow avalanches at any time of year. In mountainous terrain, avalanches are among the most serious objective natural hazards to life and property, with their destructive capability resulting from their potential to carry enormous masses of snow at high speeds.

There is no universally accepted classification of avalanches— different classifications are useful for different purposes. Avalanches can be described by their size, their destructive potential, their initiation mechanism, their composition and their dynamics.

Formation and Type

Most avalanches occur spontaneously during storms under increased load due to snowfall. The second largest cause of natural avalanches is metamorphic changes in the snowpack such as melting due to solar radiation. Other natural causes include rain, earthquakes, rockfall and icefall. Artificial triggers of avalanches include skiers, snowmobiles, and controlled explosive work.

Avalanche initiation can start at a point with only a small amount of snow moving initially; this is typical of wet snow avalanches or avalanches in dry unconsolidated snow. However, if the snow has sintered into a stiff slab overlying a weak layer then fractures can propagate very rapidly, so that a large volume of snow, that may be thousands of cubic metres, can start moving almost simultaneously.

A snowpack will fail when the load exceeds the strength. The load is straightforward; it is the weight of the snow. However, the strength of the snowpack is much more difficult to determine and is extremely heterogenous. It varies in detail with properties of the snow grains, size, density, morphology, temperature, water content; and the properties of the bonds between the grains. These properties may all metamorphose in time according to the local humidity, water vapour flux, temperature and heat flux. The top of the snowpack is also extensively influenced by incoming radiation and the local air flow. One of the aims of avalanche research is to develop and validate computer models that can describe the evolution of the seasonal snowpack over time. A complicating factor is the complex interaction of terrain and weather, which causes significant spatial and temporal variability of the depths, crystal forms, and layering of the seasonal snowpack.

Loose Snow Avalanches

Loose snow avalanches, most common in steeper terrain, often occur in freshly fallen, low-density surface snow, or in older surface snow that has been softened by strong solar radiation. In loose snow avalanches, the release usually starts at a point and the avalanche gradually widens as it travels down the slope and entrains more snow. The characteristic shape of a loose snow avalanche is usually described as resembling a teardrop. Large, loose snow avalanches may cause slab avalanches.

Slab Avalanches

Slab avalanches form frequently in snow that has been deposited, or redeposited by wind. They have the characteristic appearance of a block of snow cut out from its surroundings by fractures. Elements of slab avalanches include the following: a crown fracture at the top of the start zone, flank fractures on the sides of the start zones, and a fracture at the bottom called the staunchwall. The crown and flank fractures are vertical walls in the snow delineating the snow that was entrained in the avalanche from the snow that remained on the slope. Slabs can vary in thickness from a few centimetres to three metres. Slab avalanches account for around 90% of avalanche-related fatalities in backcountry users.

Powder Snow Avalanches

The largest avalanches form turbulent suspension currents known as powder snow avalanches or mixed avalanches. These consist of a powder cloud, which overlies a dense avalanche. They can form from any type of snow or initiation mechanism, but usually occur with fresh dry powder. They can exceed speeds of 300 km/h, and masses of 10,000,000 tonnes; their flows can travel long distances along flat valley bottoms and even uphill for short distances.

Wet Snow Avalanches

In contrast to powder snow avalanches, wet snow avalanches are a low velocity suspension of snow and water, with the flow confined to the track surface (McClung, first edition 1999, page 108). The low speed of travel is due to the friction between the sliding surface of the track and the water saturated flow. Despite the low speed of travel (~10–40 km/h), wet snow avalanches are capable of generating powerful destructive forces, due to the large mass, and density. The body of the flow of a wet snow avalanche can plough through soft snow, and can scour boulders, earth, trees, and other vegetation; leaving exposed, and often scored, ground in the avalanche track. Wet snow avalanches can be initiated from either loose snow releases, or slab releases, and only occur in snow packs that are water saturated and isothermally equilibrated to the melting point of water. The isothermal characteristic of wet snow avalanches has led to the secondary term of isothermal slides found in the literature (for example in Daffern, 1999, page 93). At temperate latitudes wet snow avalanches are frequently associated with climatic avalanche cycles at the end of the winter season, when there is significant daytime warming.

Terrain, Snowpack, Wheather

Doug Fesler and Jill Fredston developed a conceptual model of the three primary elements of avalanches: terrain, weather, and snowpack. Terrain describes the places where avalanches occur, weather describes the meteorological conditions that create the snowpack, and snowpack describes the structural characteristics of snow that make avalanche formation possible.

Terrain

Avalanche formation requires a slope where snow can accumulate, yet has enough steepness for the snow to accelerate once set in motion by the combination of mechanical failure (of the snowpack) and gravity. The angle of the slope that can hold snow, called the angle of repose, depends on a variety of factors such as crystal form and moisture content. Some forms of drier and colder snow will only stick to lower angle slopes; while wet and warm snow can bond to very steep surfaces. In particular, in coastal mountains, such as the Cordillera del Paine region of Patagonia, deep snowpacks collect on vertical, and overhanging, rock faces. The angle of slope that can allow moving snow to accelerate depends on a variety of factors such as the snow's shear strength, which is itself dependent upon crystal form, and the configuration of layers and inter-layer interfaces.

The snowpack on slopes with sunny exposures is strongly influenced by sunshine. Diurnal cycles of thawing and refreezing can stabilize the snowpack by promoting settlement. Strong freeze thaw cycles result in the formation of surface crusts during the night, and the formation of unstable surface snow during the day. Slopes in the lee of a ridge or other wind obstacle accumulate more snow and are more likely to include pockets of deep snow, wind slabs, and cornices, all of which, when disturbed, may result in avalanche formation. Conversely the snowpack on a windward slope is often much shallower than on lee slopes.

Avalanches and avalanche paths share common elements: a start zone where the avalanche originates, a track along which the avalanche flows, and a runout zone where the avalanche comes to rest. The debris deposit is the accumulated mass of the avalanched snow once it has come to rest in the runout zone. For the image at left, many small avalanches form in this avalanche path every year, but most of these avalanches do not run the full vertical or horizontal length of the path. The frequency with which avalanches form in a given area is known as the return period.

The start zone of an avalanche must be steep enough to allow snow to accelerate once set in motion, additionally convex slopes are less stable than concave slopes, because of the disparity between the tensile strength of snow layers and their compressive strength. The composition and structure of the ground surface beneath the snowpack influences the stability of the snowpack, either being a source of strength or weakness. Avalanches are unlikely to form in very thick forests, however boulders and sparsely distributed vegetation can create weak areas deep within the snowpack, through the formation of strong temperature gradients. Full-depth avalanches (avalanches that sweep a slope virtually clean of snow cover) are more common on slopes with smooth ground cover, such as grass or rock slabs.

Generally speaking, avalanches follow drainages down slope, frequently sharing drainage features with summertime watersheds. At and below tree line, avalanche paths through drainages are well defined by vegetation boundaries called trim lines, which occur where avalanches have removed trees and prevented regrowth of large vegetation. Engineered drainages, such as the avalanche dam on Mount Stephen in Kicking Horse Pass, have been constructed to protect people and property, by redirecting the flow of avalanches. Deep debris deposits from avalanches will collect in catchments at the terminus of a run out, such as gullies, and river beds.

Slopes flatter than 25 degrees or steeper than 60 degrees typically have a lower incidence of avalanche involvement. Human triggered avalanches have the greatest incidence when the snow's angle of repose is between 35 and 45 degrees; the critical angle, the angle at which human-triggered avalanches are most frequent, is 38 degrees. But when the incidence of human triggered avalanches are normalized by the rates of recreational use hazard increases uniformly with slope angle, and no significant difference in hazard for a given exposure direction can be found. The rule of thumb is: *A slope that is flat enough to hold snow but steep enough to ski has the potential to generate an avalanche, regardless of the angle.*

Snowpack Structure and Characteristics

Figure: *After surface hoarfrost becomes buried by later snowfall, the buried hoar layer can be a weak layer upon which upper layers can slide.*

The snowpack is composed of ground-parallel layers that accumulate over the winter. Each layer contains ice grains that are representative of the distinct meteorological conditions during which the snow formed and was deposited. Once deposited, a snow layer continues to evolve under the influence of the meteorological conditions that prevail after deposition.

For an avalanche to occur, it is necessary that a snowpack have a weak layer (or instability) below a slab of cohesive snow. In practice the formal mechanical and structural factors related to snowpack instability are not directly observable outside of laboratories, thus the more easily observed properties of the snow layers (e.g. penetration resistance, grain size, grain type, temperature) are used as index measurements of the mechanical properties of the snow (e.g. tensile strength, friction coefficients, shear strength, and ductile strength). This results in two principal sources of uncertainty in determining snowpack stability based on snow structure: First, both the factors influencing snow stability and the specific characteristics of the snowpack vary widely within small areas and time scales, resulting in significant difficulty extrapolating point observations of snow layers across different scales of space and time. Second, the relationship between readily observable snowpack characteristics and the snowpack's critical mechanical properties has not been completely developed.

While the deterministic relationship between snowpack characteristics and snowpack stability is still a matter of ongoing scientific study, there is a growing empirical understanding of the snow composition and deposition characteristics that influence the likelihood of an avalanche. Observation and experience has shown that newly fallen snow requires time to bond with the snow layers beneath it, especially if the new snow falls during very cold and dry conditions. If ambient air temperatures are cold enough, shallow snow above or around boulders, plants, and other discontinuities in the slope, weakens from rapid crystal growth that occurs in the presence of a critical temperature gradient. Large, angular snow crystals are indicators of weak snow, because such crystals have fewer bonds per unit volume than small, rounded crystals that pack tightly together. Consolidated snow is less likely to slough than loose powdery layers or wet isothermal snow; however, consolidated snow is a necessary condition for the occurrence of slab avalanches, and persistent instabilities within the snowpack can hide below well-consolidated surface layers. Uncertainty associated with the empirical understanding of the factors influencing snow stability leads most professional avalanche workers to recommend conservative use of avalanche terrain relative to current snowpack instability.

Weather

Avalanches can only occur in a standing snowpack. Typically winter seasons at high latitudes, high altitudes, or both, have weather that is sufficiently unsettled, and cold enough for precipitated snow to accumulate into a seasonal snowpack. Continentality, through its potentiating influence on the meteorological extremes experienced by snowpacks, is an important factor in the evolution of instabilities, and consequential the occurrence of avalanches. Conversely, proximity to coastal environments moderates the meteorological extremes experienced by snowpacks, and results in a faster stabilization of the snowpack after storm cycles. The evolution of the snowpack is critically sensitive to small variations within the narrow range of meteorological conditions that allow for the accumulation of snow into a snowpack. Among the critical factors controlling snowpack evolution are: heating by the sun, radiational cooling, vertical temperature gradients in standing snow, snowfall amounts, and snow types. Generally, mild winter weather will promote the settlement and stabilization of the snowpack; and conversely very cold, windy, or hot weather will weaken the snowpack.

At temperatures close to the freezing point of water, or during times of moderate solar radiation, a gentle freeze-thaw cycle will take place. The melting and refreezing of water in the snow strengthens the snowpack during the freezing phase and weakens it during the thawing phase. A rapid rise in temperature, to a point significantly above the freezing point of water, may cause avalanche formation at any time of year.

Persistent cold temperatures can either prevent new snow from stabilizing or destabilize the existing snowpack. Cold air temperatures on the snow surface produce a temperature gradient in the snow, because the ground temperature at the base of the snowpack is usually around °C, and the ambient air temperature can be much colder. When a temperature gradient greater than 10 °C change per vertical metre of snow is sustained for more than a day, angular crystals called depth hoar or facets begin forming in the snowpack because of rapid moisture transport along the temperature gradient. These angular crystals, which bond poorly to one another and the surrounding snow, often become a persistent weakness in the snowpack. When a slab lying on top of a persistent weakness is loaded by a force greater than the strength of the slab and persistent weak layer, the persistent weak layer can fail and generate an avalanche.

Any wind stronger than a light breeze can contribute to a rapid accumulation of snow on sheltered slopes downwind. Wind slab forms quickly and, if present, weaker snow below the slab may not have time to adjust to the new load. Even on a clear day, wind can quickly load a slope with snow by blowing snow from one place to another. Top-loading occurs when wind deposits snow from the top of a slope; cross-loading occurs when wind deposits snow parallel to the slope. When a wind blows over the top of a mountain, the leeward, or downwind, side of the mountain experiences top-loading, from the top to the bottom of that lee slope. When the wind blows across a ridge that leads up the mountain, the leeward side of the ridge is subject to cross-loading. Cross-loaded wind-slabs are usually difficult to identify visually.

Snowstorms and rainstorms are important contributors to avalanche danger. Heavy snowfall will cause instability in the existing snowpack, both because of the additional weight and because the new snow has insufficient time to bond to underlying snow layers. Rain has a similar effect. In the short-term, rain causes instability because, like a heavy snowfall, it imposes an additional load on the snowpack; and, once rainwater seeps down through the snow, it acts as a lubricant, reducing the natural friction between snow layers that holds the snowpack together. Most avalanches happen during or soon after a storm. Daytime exposure to sunlight will rapidly destabilize the upper layers of the snowpack if the sunlight is strong enough to melt the snow, thereby reducing its hardness. During clear nights, the snowpack can re-freeze when ambient air temperatures fall below freezing, through the process of long-wave radiative cooling, or both. Radiative heat loss occurs when the night air is significantly cooler than the snowpack, and the heat stored in the snow is re-radiated into the atmosphere.

Dynamics

When a slab avalanche forms, the slab disintegrates into increasingly smaller fragments as the snow travels downhill. If the fragments become small enough the outer layer of the avalanche, called a saltation layer, takes on the characteristics of a fluid. When sufficiently fine particles are present they can become airborne and, given a sufficient quantity of airborne snow, this portion of the avalanche can become separated from the bulk of the avalanche and travel a greater distance as a powder snow avalanche. Scientific studies using radar, following the 1999 Galtür avalanche disaster, confirmed the hypothesis that a saltation layer forms between the surface and the airborne components of an avalanche, which can also separate from the bulk of the avalanche.

Driving an avalanche is the component of the avalanche's weight parallel to the slope; as the avalanche progresses any unstable snow in its path will tend to become incorporated, so increasing the overall weight. This force will increase as the steepness of the slope increases, and diminish as the slope flattens. Resisting this are a number of components that are thought to interact with each other: the friction between the avalanche and the surface beneath; friction between the air and snow within the fluid; fluid-dynamic drag at the leading edge of the avalanche; shear resistance between the avalanche and the air through which it is passing, and shear resistance between the fragments within the avalanche itself. An avalanche will continue to accelerate until the resistance exceeds the forward force.

Modelling

Attempts to model avalanche behaviour date from the early 20th century, notably the work of Professor Lagotala in preparation for the

1924 Winter Olympics in Chamonix. His method was developed by A. Voellmy and popularised following the publication in 1955 of his *Ueber die Zerstoerungskraft von Lawinen* (On the Destructive Force of Avalanches).

Voellmy used a simple empirical formula, treating an avalanche as a sliding block of snow moving with a drag force that was proportional to the square of the speed of its flow:

$$\text{Pref} = \frac{1}{2}\rho v^2$$

He and others subsequently derived other formulae that take other factors into account, with the Voellmy-Salm-Gubler and the Perla-Cheng-McClung models becoming most widely used as simple tools to model flowing (as opposed to powder snow) avalanches.

Since the 1990s many more sophisticated models have been developed. In Europe much of the recent work was carried out as part of the SATSIE (Avalanche Studies and Model Validation in Europe) research project supported by the European Commission which produced the leading-edge MN2L model, now in use with the *Service Réstitution Terrains en Montagne* (Mountain Rescue Service) in France, and D2FRAM (Dynamical Two-Flow-Regime Avalanche Model), which was still undergoing validation as of 2007.

Human Involvement

Figure: *Snow fences in Switzerland*

Figure: *Avalanche blasting in French ski resort Tignes (3,600 m)*

Prevention

Preventative measures are employed in areas where avalanches pose a significant threat to people, such as ski resorts and mountain towns, roads and railways. There are several ways to prevent avalanches and lessen their power and destruction; active preventative measures reduce the likelihood and size of avalanches by disrupting the structure of the snowpack; passive measures reinforce and stabilize the snowpack in situ. The simplest active measure is by repeatedly traveling on a snowpack as snow accumulates; this can be by means of boot-packing, ski-cutting, or machine grooming. Explosives are used extensively to prevent avalanches, by triggering smaller avalanches that break down instabilities in the snowpack, and removing over burden that can result in larger avalanches. Explosive charges are delivered by a number of methods including hand tossed charges, helicopter dropped bombs, Gazcx concussion lines, and ballistic projectiles launched by air cannons and artillery. Passive preventive systems such as Snow fences and light walls can be used to direct the placement of snow. Snow builds up around the fence, especially the side that faces the prevailing winds. Downwind of the fence, snow buildup is lessened. This is caused by the loss of snow at the fence that would have been deposited and the pickup of the snow that is already there by the wind, which was depleted of snow at the fence. When there is a sufficient density of trees, they can greatly reduce the strength of avalanches. They hold snow in place and when there is an avalanche, the impact of the snow against the trees slows it down. Trees can either be planted or they can be conserved, such as in the building of a ski resort, to reduce the strength of avalanches.

Mitigation

In many areas, regular avalanche tracks can be identified and precautions can be taken to minimise damage, such as the prevention of development in these areas. To mitigate the effect of avalanches the construction of artificial barriers can be very effective in reducing avalanche damage. There are several types: One kind of barrier (snow net) uses a net strung between poles that are anchored by guy wires in addition to their foundations. These barriers are similar to those used for rockslides. Another type of barrier is a rigid fence-like structure (snow fence) and may be constructed of steel, wood or pre-stressed concrete. They usually have gaps between the beams and are built perpendicular to the slope, with reinforcing beams on the downhill side. Rigid barriers are often considered unsightly, especially when many rows must be built. They are also expensive and vulnerable to damage from falling rocks in the warmer months. In addition to industrially manufactured barriers, landscaped barriers, called avalanche dams stop or deflect avalanches with their weight and strength. These barriers are made out of concrete, rocks or earth. They are usually placed right above the structure, road or railway that they are trying to protect, although they can also be used to channel avalanches into other barriers. Occasionally, earth mounds are placed in the avalanche's path to slow it down. Finally, along transportation corridors, large shelters, called snow sheds, can be built directly in the slide path of an avalanche to protect traffic from avalanches.

Survival, Rescue, and Recovery

Avalanche accidents are broadly differentiated into 2 categories: accidents in recreational settings, and accidents in residential, industrial, and transportation settings. This distinction is motivated by the observed difference in the causes of avalanche accidents in the two settings. In the recreational setting most accidents are caused by the people involved in the avalanche. In a 1996 study, Jamieson et al. (pages 7–20) found that 83% of all avalanches in the recreational setting were caused by those who were involved in the accident. In contrast, all of the accidents in the residential, industrial, and transportation settings were due to spontaneous natural avalanches. Because of the difference in the causes of avalanche accidents, and the activities pursued in the two settings, avalanche and disaster management professionals have developed two related preparedness, rescue, and recovery strategies for each of the settings.

Notable Avalanches

Two avalanches occurred in March 1910 in the Cascade and Selkirk Mountain ranges; On March 1 the Wellington avalanche killed 96 in Washington State, United States. Three days later 62 railroad workers were killed in the Rogers Pass avalanche in British Columbia, Canada.

During World War I, an estimated 40,000 to 80,000 soldiers died as a result of avalanches during the mountain campaign in the Alps at the Austrian-Italian front, many of which were caused by artillery fire. Some 10,000 men, from both sides, lost their lives in avalanches in December 1916. However, it is very doubtful avalanches were used deliberately at the tactical level as weapons; more likely they were simply a side effect to shelling enemy troops, occasionally adding to the toll taken by the artillery. Avalanche prediction is nearly impossible; forecasters can only assert the conditions, terrain and relative likelihood of slides with the help of detailed weather reports and from localized snowpack observation. It would be almost impossible to predict avalanche conditions many miles behind enemy lines, making it impossible to intentionally target a slope at risk for avalanches. Also, high priority targets received continual shelling and would be unable to build up enough unstable snow to form devastating avalanches, effectively imitating the avalanche prevention programmes at ski resorts.

In the northern hemisphere winter of 1950–1951 approximately 649 avalanches were recorded in a three-month period throughout the Alps in Austria, France, Switzerland, Italy and Germany. This series of avalanches killed around 265 humans and was termed the Winter of Terror.

A mountain climbing camp on Lenin Peak, in what is now Kyrgyzstan, was wiped out in 1990 when an earthquake triggered a large avalanche that overran the camp. Forty-three climbers were killed.

In 1993, the Bayburt Üzengili avalanche killed 60 individuals in Üzengili in the province of Bayburt, Turkey.

A large avalanche in Montroc, France, in 1999, 300,000 cubic metres of snow slid on a 30° slope, achieving a speed of 100 km/h (60 mph). It killed 12 people in their chalets under 100,000 tons of snow, 5 metres (15 ft) deep. The mayor of Chamonix was convicted of second-degree murder for not evacuating the area, but received a suspended sentence.

The small Austrian village of Galtür was hit by the Galtür avalanche in 1999. The village was thought to be in a safe zone but the avalanche was exceptionally large and flowed into the village. Thirty-one people died.

An avalanche in the Siachen glacier in the Himalaya mountains buried at least 124 Pakistani soldiers and 11 civilians in April 2012.

An avalanche on Manaslu, in the Himalaya mountains, overwhelmed camp 3, early in morning of September 23, 2012 killing 11 climbers, and skiers.

Classification

European avalanche risk table:

In Europe, the avalanche risk is widely rated on the following scale, which was adopted in April 1993 to replace the earlier non-standard national schemes. Descriptions were last updated in May 2003 to enhance uniformity.

In France, most avalanche deaths occur at risk levels 3 and 4. In Switzerland most occur at levels 2 and 3. It is thought that this may be due to national differences of interpretation when assessing the risks.

Risk Level	*Snow Stability*	*Avalanche Risk*
1 - Low	Snow is generally very stable.	Avalanches are unlikely except when heavy loads are applied on a very few extreme steep slopes. Any spontaneous avalanches will be minor sloughs. In general, safe conditions.
2 - Limited	On some steep slopes the snow is only moderately stable. Elsewhere it is very stable.	Avalanches may be triggered when heavy loads are applied, especially on a few generally identified steep slopes. Large spontaneous avalanches are not expected.
3 - Medium	On many steep slopes the snow is only moderately or weakly stable.	Avalanches may be triggered on many slopes even if only light loads are applied. On some slopes, medium or even fairly large spontaneous avalanches may occur.
4 - High	On most steep slopes the snow is not very stable.	Avalanches are likely to be triggered on many slopes even if only light loads are applied. In some places, many medium or sometimes large spontaneous avalanches are likely.
5 - Very High	The snow is generally unstable.	Even on gentle slopes, many large spontaneous avalanches are likely to occur.

Stability:

- Generally described in more detail in the avalanche bulletin (regarding the altitude, aspect, type of terrain etc.)

additional load:

- heavy: two or more skiers or boarders without spacing between them, a single hiker or climber, a grooming machine, avalanche blasting
- light: a single skier or snowboarder smoothly linking turns and without falling, a group of skiers or snowboarders with a minimum 10 m gap between each person, a single person on snowshoes

Gradient:

- gentle slopes: with an incline below about 30°
- steep slopes: with an incline over 30°
- very steep slopes: with an incline over 35°
- extremely steep slopes: extreme in terms of the incline (over 40°), the terrain profile, proximity of the ridge, smoothness of underlying ground

European Avalanche Size Table

Avalanche size:

Size	*Runout*	*Potential Damage*	*Physical Size*
1 - Sluff	Small snow slide that cannot bury a person, though there is a danger of falling.	Unlikely, but possible risk of injury or death to people.	length <50 m volume <100 m^3
2 - Small	Stops within the slope.	Could bury, injure or kill a person.	length <100 m volume <1,000 m^3
3 - Medium	Runs to the bottom of the slope.	Could bury and destroy a car, damage a truck, destroy small buildings or break trees.	length <1,000 m volume <10,000 m^3
4 - Large	Runs over flat areas (significantly less than 30°) of at least 50 m in length, may reach the valley bottom.	Could bury and destroy large trucks and trains, large buildings and forested areas.	length >1,000 m volume >10,000 m^3

North American Avalanche Danger Scale

In the United States and Canada, the following avalanche danger scale is used. Descriptors vary depending on country.

Canadian Classification for Avalanche Size

The Canadian classification for avalanche size is based upon the consequences of the avalanche. Half sizes are commonly used.

Size	*Destructive Potential*
1	Relatively harmless to people.
2	Could bury, injure or kill a person.
3	Could bury and destroy a car, damage a truck, destroy a small building or break a few trees.
4	Could destroy a railway car, large truck, several buildings or a forest area up to 4 hectares.
5	Largest snow avalanche known. Could destroy a village or a forest of 40 hectares.

United States Classification for Avalanche Size

Size	*Destructive Potential*
1	Sluff or snow that slides less than 50m (150') of slope distance.
2	Small, relative to path.
3	Medium, relative to path.
4	Large, relative to path.
5	Major or maximum, relative to path.

Rutschblock Test

Slab avalanche hazard analysis can be done using the Rutschblock Test. A 2 m wide block of snow is isolated from the rest of the slope and progressively loaded. The result is a rating of slope stability on a seven step scale.

Avalanche Control

Avalanche control or avalanche defence activities reduce the hazard avalanches pose to human life, activity, and property. Avalanche control begins with a risk assessment conducted by surveying for potential avalanche terrain by identifying geographic features such as vegetation patterns, drainages, and seasonal snow distribution that are indicative of avalanches. From the identified avalanche risks, the hazard is assessed by identifying threatened human geographic features such as roads, ski hills, and buildings. Avalanche control programmes address the avalanche hazard by formulating prevention and mitigation plans, which are then executed during the winter season. The prevention and mitigation plans combine extensive snow pack observation with three major groups of interventions: active, passive, and social sometimes more narrowly defined as "explosive", "structural", and "awareness", according to the prevalent techniques used. Avalanche control techniques either directly intervene in the evolution of the snow pack, or lessen the effect of an avalanche once it has occurred. For the event of human involvement, avalanche control organizations develop and train exhaustive response and recovery plans.

Figure: *An avalanche snow bridge near a ski-resort in Vorarlberg.*

Risk and Hazard Assessment

Risk assessment geographical surveys for potential avalanche terrain by studying topography, vegetation, and seasonal snow distribution. Hazard assessment geographical surveys for the consequences of avalanche by studying exposure of urbanization, industrialization, transportation, recreational activities, and the distribution of human use of the potential avalanche terrain identified in the risk assessment.

Prevention and Mitigation

Prevention and mitigation begins with observing the snow pack to forecast the risk of avalanche occurrence. The forecast risk then determines the necessary interventions to reduce the hazard posed by an avalanche.

Observation and Forecasting

Snow pack observation studies the layering and distribution of the snow to estimate the instabilities of the snow pack and thus the risk of an avalanche occurring in a particular terrain feature. In areas of heavy human use the snow pack is monitored throughout the winter season to assess its evolution under the prevailing meteorological conditions. In contrast to heavily used avalanche terrain where forecasting is the goal of snow observation, in remote terrain, or terrain that is infrequently visited, snow pack observation elucidates the immediate instabilities of the snow pack. We must be aware and be alert during the time of avalanche.

Active Interventions

Active techniques reduce the risk of an avalanche occurring by promoting the stabilization and settlement of the snow pack through three forms of intervention: disrupting weak layers in the snow pack, increasing the uniformity of the snow pack, and lessening the amount of snow available in snow pack for entrainment in an avalanche; this can be accomplished either by triggering smaller less hazardous avalanches, or by directly influencing the structure of the layering of the snow pack. Active avalanche control can be broadly classified into either mechanical or explosive methods. Mechanical methods are typically used in either remote terrain, smaller terrain, or less hazardous terrain; while explosive methods are used in accessible large high hazard terrain, or terrain with industrial, commercial recreational, urbanized, and transportation usage.

In the smallest terrain features the simplest method of avalanche control is a mechanical intervention that disrupts weak snow layers by directly walking through them, a technique referred to as boot packing. For larger features this method can be extended by mechanized redistribution of snow using large tracked vehicles called snow groomers. These two mechanical interventions can only be safely done as the snow is deposited and before it develops any instabilities. In terrain that can only be sporadically accessed, or in a highly developed snow pack that is too deep for boot packing, ski stabilization techniques are used. The first technique of ski stabilizing is a method of entering a slope called ski cutting. In this method a skier attempts to trigger a small avalanche by breaking the tensile support of the upper snow pack through a quick traverse along the top of the slope, the skier can be belayed on a rope to further protect them from being caught in an avalanche. A snow pack can then be further settled out, or stabilized, by further down slope ski traffic through it. Finally knotted cord can be used to saw through the roots of cornices, causing the cornice to drop onto the snow pack of the slope below. This has the combined effect of reducing the objective hazard posed by the cornice, and providing a large impact force on the snow pack.

Explosive techniques involve the artificial triggering of smaller less destructive avalanches, by detonating charges either above or on the snow surface. The explosives may be deployed by manually hand tossing and lowering, by bombing from a helicopter, or by shelling with a small howitzer, recoilless rifle, or air gun. In balancing the hazard to personnel with the effectiveness of the deployment method at accessing and triggering avalanche terrain, each method has its drawbacks and advantages. Among the newest methods, strategically placed remote controlled installations that generate an air blast by detonating a fuel-air explosive above the snow pack in an avalanche starting zone, offer fast and effective response to avalanche control decisions while minimizing the risk to avalanche control personnel; a feature especially important for avalanche control in transportation corridors. For example the Avalanche Towers (Sprengmast) installed in Switzerland, Austria, and Norway use solar powered launchers to deploy charges from a magazine containing 12 radio controlled charges. The magazines can be transported, loaded, and removed from the towers by helicopter, without the need for a flight assistant, or on site personnel.

Explosive control has proved to be effective in areas with easy access to avalanche starting areas and where minor avalanches can be tolerated. It is mostly unacceptable, however, in areas with human residence and where even a small probability of a larger avalanche is probable,

Permanent Interventions

Permanent techniques slow, stop, divert, or prevent snow from moving; either completely or to enough of an extent that the destructive forces are significantly lessened. Permanent techniques involve constructing structures and modifying terrain for purposes classified as:

- Snow retention structures (snow racks, avalanche snow bridges, snow nets), used in the upper path of probable avalanche paths
- Avalanche barriers: The main part of the avalanche barriers is based on a high tensile strength steel wire mesh, extending across the slope and reaching to the surface of the snow. The supporting effect created by the retaining surface prevents possible creeping within the snow cover and sliding of the snow cover on the terrain surface. Breaking-away of avalanches is

thus prevented at the starting zone, while occurring snow movements are restricted to the extent that they remain harmless. The forces resulting from the snow pressure are absorbed by the snow nets and carried off over the swivel posts and anchor ropes into the anchor points.

- Snow guard devices (used to increase snow retention on roofs).
- Snow redistribution structures (wind baffles, snow fences)
- Snow deflection structures used to deflect and confine the moving snow within the avalanche track.. They should not deflect the avalanche sharply, because in the latter case they may be easily overrun by snow.
- Snow retardation structures (e.g. snow breakers), mostly used in small-slope parts of the avalanche track, to enhance the natural retardation
- Snow catchment structures
- Direct protection of important objects and structures, e.g., by snow sheds (avalanche sheds).

A single intervention may fulfill the needs of multiple classes of purpose, for example Avalanche dams, ditches, earth mounds, and terraces are used for deflection, retardation, and catchment. Other passive methods include:

- reforestation, up the natural tree line — forests serve all the functions of artificial avalanche defences: retention, redistribution, retardation and catchment
- snow caves, as well as recessed, dug out, and snow walled quinzhees and bivouac shelters are used to temporarily protect bivouacking climbers and skiers by providing them with breathing space in the event of burial by avalanches.
- Architectural streamlining and wedge shaping buildings, such as those found in the historic high mountain villages of the Alps..

Social Interventions

To mitigate the hazard of avalanches, social interventions reduce the incidence and prevalence of human avalanche involvement by modifying the behaviour of people, so that their use of avalanche terrain is adapted to prevent their involvement in avalanches. Avalanche control organizations accomplish this by targeting awareness and education programmes at communities that frequent avalanche terrain. Surveys of avalanche accidents have observed that most avalanches that involve people are caused by people, and of those victims many were unaware of the risk of avalanche occurrence. To address this observation, introductory awareness and education programmes provide instruction in the avoidance of hazardous avalanche involvement through the recognition of avalanche terrain, the observation of snow pack instabilities, and the identification of human activities that cause avalanches. Avalanche control organizations also publicly disseminate forecasts, bulletins, warnings, and reports of avalanche activity to assist communities of avalanche terrain users.

Response and Recovery

Avalanche control organizations plan for, and respond, to avalanches. Typical responses span from clearing transportation corridors of avalanche debris, to repairing industrial and recreational facilities, to search, rescue, and recovery. To improve the outcome of human avalanche involvement avalanche control organizations offer training and education to both professionals and recreational amateurs in avalanche preparedness.

Professional Preparedness

Professional responses to avalanches are targeted at avalanches involving the general unprepared public. When avalanches are forecast to occur, avalanche terrain to which the general unprepared public is exposed will be closed, and after the avalanches have occurred the area is cleared of debris, and repaired. When unexpected avalanches occur that involve the general unprepared public, avalanche control organizations respond with large professionally organized search teams involving probe lines, and trained search and rescue dogs.

Amateur Preparedness

Recreational response to avalanches involves the rapid formation of an ad hoc search and rescue team. The ad hoc search and rescue teams rely on all the participants having prepared for a potential avalanche by carrying the correct search and rescue equipment, and undergoing the appropriate training.

Mudslide

A mudslide is the most rapid (up to 80 km/h, or 50 mph) and fluid type of downhill mass wasting. It is a rapid movement of a large mass of mud formed from loose soil and water. Similar phenomena include mudflow, mud stream, debris flow (e.g. in high mountains), jökulhlaup, and lahar (from volcanoes). These terms cover a broad variety in water content (from loamy mud, to almost liquid, and even steam); material (wet soil, sandysediments and/or silt, dirt, rock, debris, volcanic ash, small plants, and even trees); length; total mass; and velocity.

Triggering of Mudslides

Figure: *The Mameyes mudslide disaster, in barrio Tibes, Ponce, Puerto Rico, was caused by heavy rainfall from Tropical Storm Isabel in 1985. The mudslide destroyed more than 100 homes and claimed an estimated 300 lives.*

Heavy rainfall, snowmelt, or high levels of ground water flowing through cracked bedrock may trigger a movement of soil or sediments. Floods, debris- and mud flows may also occur when strong rains on hill or mountain slopes cause extensive erosion and/or what is known as "channel scour". The 2006 Sidoarjo mud flow may have been caused by rogue drilling.

Some broad mudflows are rather viscous and therefore slow; others begin very quickly and continue like an avalanche. If large enough they can devastate villages and countrysides. They are composed of at least 50% silt and clay-sized materials and up to 30% water. Mudflows are common even in the hills around Los Angeles where they have destroyed many homes built on hillsides without sufficient support.

The point where a muddy material begins to flow depends on its grain and the water content. Fine grainy material or soil has a smaller friction angle than a coarse sediment or a debris flow, but falling rock pieces can trigger a material flow, too. On December 14, 1999 in Vargas, Venezuela, a mudslide known as The Vargas tragedy, which significantly altered more than 60 kilometres (37 mi) of its coastline, was triggered due to heavy rainfall and caused estimated damages of USD $1.79 to $3.5 billion, a death toll considered to be between 10,000 and 30,000, 85,000 people evacuated and led to the complete collapse of the state's infrastructure.

Mudslides and Landslides

Landslide is a general term referring to all types of surface movement particularly avalanches involving the mass movement downhill of soil, rock or snow under the action of gravity. Landslides and avalanches can be devastating in terms of human life and general destruction. Mudslides are caused by unusually heavy rain or a sudden thaw. They consist mainly of mud and water plus fragments of rock and other debris, this causes them to behave like a flood. They can move houses off their foundations or bury a place within minutes, this is due to their incredibly strong currents. Every year combined they kill more people than typhoons, hurricanes and cyclones do.

Mudslide Geography

When a mudslide occurs it is given four named areas, the 'main scarp', in bigger mudslides the 'upper and lower shelves' and the 'toe'. The main scarp will be the original area of incidence, the toe is the last affected area(s). The upper and lower shelves are located wherever a large dip (due to mountain or natural drop) in the mudslides path. Due to this a mudslide can have many shelves.

Largest Recorded Mudslide

The world's largest historic landslide (in terms of volume) occurred during the 1980 eruption of Mount St. Helens, a volcano in the Cascade Mountain Range in the State of Washington, USA. The volume of material was 2.8 km^3. Directly in the path of the huge mudslide was Spirit Lake. Normally a chilly 5°C, the mudslide instantly raised the temperature to near 38°C. Today the bottom of Spirit Lake is 100 feet above the original surface. The lake now has two and a half times more surface area than it did before the eruption.

However, the world's largest prehistoric landslide, discovered to date, took place in south-western Iran, and is named the Saidmarreh Landslide. The landslide was located on the KabirKuh anticline in Southwest Iran at 33 degrees N, 47.65 degrees E. The landslide had a volume of about 20 cubic kilometres, a depth of 300 m, a travel distance of 14 km and a width of 5 km. This means that about 50 billion tons of rock moved in this single event

The largest prehistoric landslide including submarine ones was an enormous submarine landslide that disintegrated 60,000 years ago produced the longest flow of sand and mud yet documented on Earth. The massive submarine flow travelled 1,500 kilometres – the distance from London to Rome – before depositing its load.

Areas at Risk

The area most generally recognized as being at risk of a dangerous mudslide are:

- Areas where wildfires or human modification of the land have destroyed vegetation
- Areas where landslides have occurred before
- Steep slopes and areas at the bottom of slopes or canyons
- Slopes that have been altered for construction of buildings and roads
- Channels along a stream or river
- Areas where surface runoff is directed

Protection against Mudslides

Before intense rainfall or storms, check to see if your area has had mudslides in the past, and educate yourself regarding local area evacuation or emergency plans. Families should also consider making their own emergency plan, and may wish to consider permanently leaving an area where mudslides are frequent.

During intense rainfall or storms, listen to weather warnings on the radio, look for tilted trees or posts, and listen for rumbling sounds which might indicate that a landslide or mudslide is approaching. If you see a small trickle of mud in a wash, creek or stream, move away from the area as quickly as possible; it may be a warning sign that a larger flow is imminent. Finally, if you do see a mudslide approaching, you should get to high ground out of its path as quickly as possible. If you see rocks or other debris approaching your position, find shelter and take cover, if possible under a piece of heavy furniture such as a desk or table.

After a mudslide has occurred, check for injured or trapped people, but try to avoid entering the mudslide's path. Listen for emergency instructions on the radio or TV. If you see that utility lines or cables in the area are broken, report them to the proper authorities as soon as possible.

Landslide Dam

A landslide dam, debris dam, or barrier lake is a natural damming of a river by some kind of mass wasting: landslide, debris flow, rock avalanche or volcano. If it is caused by earthquake, it may also be called a quake lake. Some landslide dams are as high as the largest existing artificial dam.

Causes

The major causes for landslide dams investigated by 1986 are landslides from excessive precipitation and earthquakes, which account for 84%. Volcanic eruptions account for a further 7% of dams. Other causes of landslides account for the remaining 9%.

Consequences

The water impounded by a landslide dam may create a dam reservoir (lake) that may last from short times to several thousand years. Because of their rather loose nature and absence of controlled spillway, landslide dams frequently fail catastropically and lead to downstream flooding, often with high casualties. A common failure scenario is overflowing with subsequent dam breach and erosion by the overflow stream.

Landslide dams are responsible for two types of flooding: backflooding (upstream flooding) upon creation and downstream flooding upon failure. Compared with catastrophic downflooding, relative slow backflooding typically presents little life hazard, but property damage can be substantial.

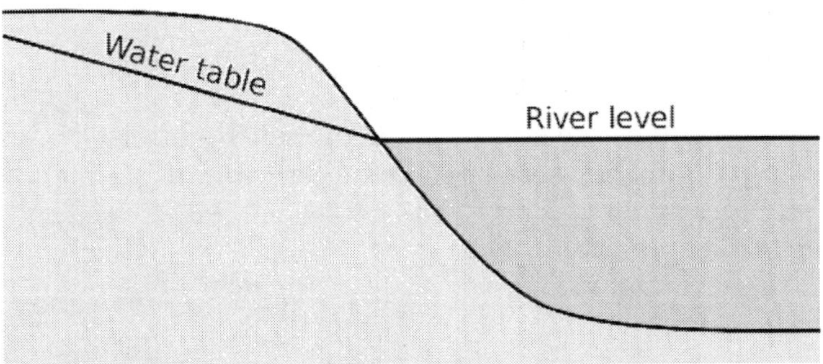

Figure: *Profiles of the dam reservoir and groundwater upstream (the landslide dam is not shown in the figure)*

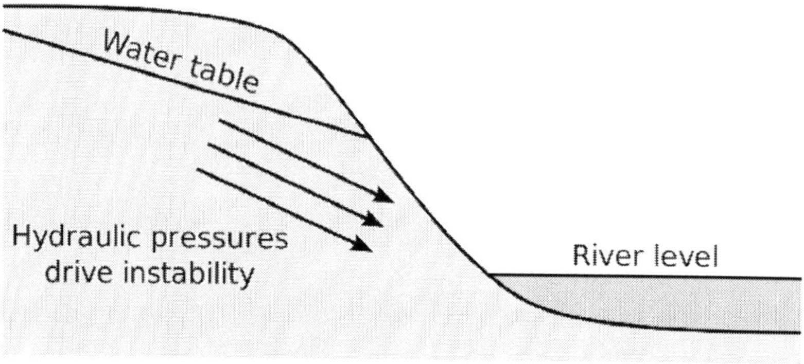

Figure: *Groundwater after dam failure downstream*

While the dam is being filled, the surrounding groundwater level rises. The dam failure may trigger further catastrophic processes. As the water level rapidly drops, the uncompensated groundwater hydraulic pressure may initiate additional landslides. Those that fall into the dam reservoir may lead to further catastrophic spillages. Moreover, the resulting flood may undercut the sides of the river valley to further produce landslides downstream.

After forming, the dam leads to aggradation of the valley upstream, and dam failure leads to aggradation downstream.

Construction engineers responsible for design of artificial dams and other structures in river valleys must take into account the potential of such events leading to abrupt changes in river's regimen.

Examples

- The highest known landslide dam of historic times is the Usoi Dam in modern Tajikistan created by a landslide triggered by an earthquake on February 18, 1911. It dammed the Murghab River to the height of 301 m (987 ft) to impound Sarez Lake 932 ft (284 m) deep.
- Lake Waikaremoana in New Zealand was formed by a 250m high landslide dam believed to be 2,200 years old. Between 1935 and 1950 the landslide was tunneled and sealed to stabilise it so it could be used for hydroelectric power generation. This appears to be the first example of modification of a natural landslide dam for power generation.
- Attabad Lake in Pakistan was formed by a landslide in 2010. (100m high)
- The "Red Lake" (Romanian: LaculRo°u) is a barrier lake in the Eastern Carpathians chain in Harghita County, Romania. The

name of "LaculRo°u" comes from the reddish alluvia deposited in the lake by the Red Creek.

- The "GrosVentre landslide" is an example of a huge, short-lived and disastrous landslide dam (60 m high).
- Quake Lake, created in Montana in 1959. (58m high)
- The Tangjiashan Lake, a dangerous "quake lake", was created as a result of the 2008 Sichuan earthquake. It was located in the extremely rugged terrain of Tangjiashan Mountain. Chinese engineers, scientists, and soldiers were involved in the digging of a sluice to alleviate the dangers of this one of 34 landslide dams created by the magnitude 8.0 Sichuan earthquake. On June 10, 2008, the lake started to drain via the sluice, flooding the evacuated town of Beichuan. (52m high)
- The Riñihuazo begun on 22 May 1960, after a landslide caused by the Great Chilean Earthquake blocked Riñihue Lake's outflow. According to the chronicler Mariño de Lobera a similar event occurred after the 1575 Valdivia earthquake. (26m high)
- Among the most destructive landslide lake outburst floods in recorded history occurred in the Sichuan province of China on 10 June 1786 when the dam on the Dadu River burst, causing a flood that extended 1400km downstream and killed 100,000 people.
- The largest landslide dam in Britain impounds Tal-y-llyn Lake (LlynMwyngil) beside CadairIdris in Snowdonia National Park, Wales. It was first recognised by E Watson in 1962. It is probably over 10,000 years old.

A Brief Overview of Landslide Monitoring

Landslide monitoring is required for a wide variety of reasons. These may include; to determine the extent, magnitude and style of landslide movement, for risk and even emergency risk management assessments and/or to assist with the design and implementation of site remedial and/or mitigation works.

Landslide monitoring can be undertaken in many forms. The monitoring may include traditional ground survey techniques repeated over time to determine movement at specific locations. It may include subsurface investigations with the aid of borehole inclinometres and their periodic manual monitoring and even continuous automated monitoring. Over recent years, automated robotic ground survey using various laser scanning techniques have been shown to be cost effective in certain applications. Even satellite based radar interferometry techniques over large areas at moderately high grid resolutions, with movements being detected down to cm level resolutions are being applied.

Inclinometre Monitoring

Where subsurface investigations are being carried out a widely used technique internationally involves the installation of inclinometre casings in boreholes and subsequent monitoring of the casings (typically installed vertically) using manual inclinometre probes.

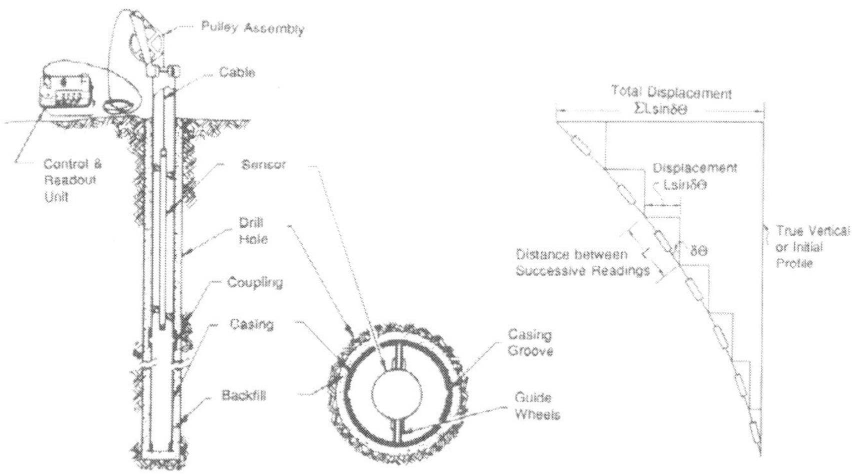

Figure: *Principle of Inclinometre operations, after Green and Mikkelsen, 1988.*

With time, repeated visits to the site whereby full length casing profiles can be recorded, the displacement history of the full subsurface profile can be determined. In Figure depicted earlier, the ground surface is shown as the top most horizontal line and depth below the ground surface is shown on the left hand y axis in metres. Horizontal displacement of the ABS plastic inclinometre casing is shown in the bottom horizontal x axis in millimetres. The data shown in figure given has been recorded at a landslide in the northern suburbs of Wollongong and the borehole drilling at this location shows bedrock occurs at a depth of approximately 8.2m. The inclinometre profiles clearly show landslide subsurface shear displacement occurring at a depth of between 7m and 8m below ground level with approximately 150mm of shear occurring at a depth of 7m below the ground surface. The first profile shown in this graph (the vertical y axis) was recorded on the 9th March 1989, and the last profile shown, was recorded on the 15th May 1996.

Such manual inclinometre profiling techniques have been employed in the Wollongong area for up to three decades. Since the mid 1990's, the UoW LRT has employed Coffey Geotechnics to conduct inclinometre monitoring at a number of landslide locations around the Illawarra area.

Monitoring of Pore Water Pressure

Ground water and pore water pressure plays an important role in landsliding. Rising pore water pressures, often (but not always) the result of rainfall, are typically the triggering mechanism for landsliding in the Illawarra region. Man made effects are another common triggering mechanism. Therefore, monitoring ground water pressure is an important component of any landslide monitoring strategy.

There are many ways to measure ground water and specifically pore water pressures. Currently the UoW LRT uses vibrating wire piezometres installed in sealed and isolated narrow sand filled intake zones at or near the landslide surface to monitor pore water pressures.

Rainfall Monitoring

Due to its influence of ground water and pore water pressure, rainfall is an important component of any landslide monitoring strategy. The UoW LRT currently employs tipping bucket style Pluviometres to monitor rainfall.

Monitoring History

The manually recorded monitoring history from a Scarborough landslide is shown earlier. Daily rainfall totals from of Bureau of Metreology daily rain gauge is shown along the base plot as a histogram. The cyclic periods of wet and dry are shown with the cumulative rainfall curves which total daily rainfall for the preceding 7, 30, 60, 90 and 120 day periods. Overlying this are two curves of total cumulative shear displacement as measured in two different borehole inclinometres. Above that is the really interesting section of this graph which shows the rate of landslide shear, the actual manually monitored rate of landslide movement.

This style of monitoring was discussed in detail by Flentje during his PhD research. Flentje's PhD thesis (particularly in Chapters 8 and 9) can be downloaded here.

Whilst such monitoring is very useful in geotechnical engineering and engineering geological assessment of landslides, several issues are very clear. Manual monitoring is expensive, and is only undertaken on an 'as needed' basis. The stepped nature of the cumulative displacement curve in figure given reflects not only the timing of landslide movement, but also the timing of landslide monitoring. Provided the monitoring occurs frequently, periods of movement are detected. If the monitoring is undertaken at low frequencies of say, annually, or even less frequently, episodes of movement may be lost in the 'averaging' effect of the long return period of the monitoring. Furthermore, such manual monitoring is not 'pro-active', it only tells you what has happened and requires some considerable reporting effort to get the information to management.

Continuous Landslide Monitoring

Following a long period of manual monitoring of landslides in the Wollongong region, the UoW LRT has undertaken the development of a network of Continuous Real-Time Monitoring (CRTM) landslide field stations.

Using all of the afore mentioned techniques, we have simply replaced manual inclinometres, with In Place Inclinometres (IPI's) which actually hang in borehole casings, at designated depths where manual inclinometre profiling has confirmed shear planes, or in the absence of confirmed shear planes, where it is assumed the shear planes have developed. For example, we would position an IPI to monitor shear displacement over the interval 7 to 8m as we know the style of landslide displacement in this borehole and we would expect this style of displacement to continue.

Using solar power and batteries, we have installed data loggers out on various important landslide sites and connected to the loggers a series of IPI's, vibrating wire piezometres, extensometres, tilt metres and rain gauges to monitor various aspects of landslide and other structural movement.

The continuously monitored history of a landslide in the northern Wollongong suburbs is shown in figure depicted earlier. The period covered by this graph is late 2003 to early 2005 and the data shown is recorded hourly. Figure shown earlier, includes a daily rainfall histogram along the bottom which shows two events with daily totals exceeding 80mm per day in April and October 2004. Over this period the In Place Inclinometres have provided excellent data. The IPI rate of shear displacement curves clearly displays two prominent spikes of accelerated displacement commencing on the 4th April and the 21st October 2004. The movement event, which commenced on The 4th April, continued for 5 days and peaked at 2.4mm per day. This was triggered by rainfall of 110mm and 106mm on consecutive days. The movement event, which commenced on the 21st October 2004, lasted for 7 days and peaked at 2.7mm per day on the second day. A maximum daily rainfall of 81.5mm and several other days of 15mm to 30mm triggered this short duration of movement.

The vibrating wire piezometres (vwp) have both shown quite strong responses to the rainfall.

This style of monitoring data is available in near real-time (subject only to the on site logging interval) via a secure password protected web-interface developed by the UoW LRT and the UoW CEDIR group. The site is password protected as it is a research application at this stage and it is constantly evolving.

The information being collected is enhancing our understanding of landslide and triggering mechanisms. The information is greatly facilitating the quantitative assessment of landslide frequency.

3

Landslide Hazard Assessment

In 1974, one of the largest landslides in recorded history occurred in the Mantaro River valley in the Andes Mountains of Peru (Hutchinson and Kogan, 1975). A temporary lake was formed when the slide dammed the Mantaro River, resulting in the flooding of farms, three bridges, and twelve miles of roadway. Almost five hundred people in and around the village of Mayunmarca lost their lives. This disaster is one example of the destructive potential of landslides and why they are considered hazards. While not every landslide results in catastrophe, the damage from many small ones may equal or exceed the impact of a single major failure. Thus, both large and small landslides are capable of causing significant damage and loss of life.

Assessing relative landslide hazard is the objective of the method described in this chapter. Its primary product, a landslide hazard map, provides planners with a practical and cost-effective way to zone areas susceptible to landsliding.

This method can be used both by planners and by landslide technical specialists. The planner will gain a working knowledge of concepts and considerations for incorporating landslide hazard evaluation into the planning process, using a level of evaluation suitable for each stage of the process, and thus should be able to ask the appropriate questions of the technical specialist and prepare terms of reference to ensure that the needed information is obtained. The technical specialist will find a review of landslide hazard issues and guidelines for conducting a landslide zonation. As is often the case in natural hazards management, planning studies are often the link between scientific information and the general development planning process.

The method presented, one of several that are available, has the following characteristics:

- Various thematic maps and remote sensing information usually available to a development study are used.
- It is designed to provide landslide hazard information appropriate for each of the stages of the planning process.
- Relative susceptibility to landsliding is used as a measure of the potential hazard within an area.
- It is applicable to regions with different geomorphologic and vegetation characteristics.
- It can usually be used within the time and budget constraints of a planning study.

Landslide Hazard Mapping and the Development Planning Process

The susceptibility of a given area to landslides can be determined and depicted using hazard zonation. A landslide hazard map can be prepared early in the planning study and developed in more detail as the study progresses. It can be used as a tool to help identify land areas best suited for development by examining the potential risk of landsliding. Furthermore, once landslide susceptibility is identified, investment projects can be developed which avoid, prevent, or substantially mitigate the hazard.

Determining the extent of landslide hazard requires identifying those areas which could be affected by a damaging landslide and assessing the probability of the landslide occurring within some time period. In general, however, specifying a time frame for the occurrence of a landslide is difficult to determine even under ideal conditions. As a result, landslide hazard is often represented by landslide susceptibility (Brabb, 1985).

Similar to the concept of flood-prone areas, landslide susceptibility only identifies areas potentially affected and does not imply a time frame when a landslide might occur. To simplify these concepts, landslide susceptibility will be referred to as landslide hazard in this chapter. Comparing the location of an area of proposed development to the degree of landslide hazard present enables the planner to estimate the landslide risk. This can be used to define land use capability and identify appropriate mitigation measures.

Definitions

- Landslide Hazard: as represented by susceptibility, which is the likelihood of a potentially damaging landslide occurring within a given area.
- Vulnerability: the level of population, property, economic activity, including public services, etc., at risk in a given area resulting from the occurrence of a landslide of a given type.
- Risk (specific): the expected degree of loss due to a particular landslide phenomenon.

A landslide hazard map which identifies areas of differing landslide potential may be generated. The need for such landslide hazard information may vary according to the future land use. The degree of landslide hazard present is considered relative since it represents the expectation of future landslide occurrence based on the conditions of that particular area. Another area may appear similar but, in fact, may have a differing landslide hazard due to a slightly different combination of landslide conditions. Thus, landslide susceptibility is relative to the conditions of each specific area, and it cannot be assumed to be identical for a similar appearing area.

Even with detailed investigation and monitoring, it is extremely difficult to predict landslide hazards in absolute terms. Sufficient understanding of landslide processes does exist, however, to be able to make an estimation of landslide hazard potential. The planner can use this estimation to make certain decisions regarding site suitability, type of development, and appropriate mitigation measures. Thus, the planner is determining acceptable risk.

Determining Acceptable Risk

Determining whether there is a need for landslide hazard information is the first step in ensuring that landslide risk does not exceed an acceptable level in planning future land use. The objective of landslide information is to identify which relatively landslide-susceptible areas are best suited for what types of development activities. For example, assessing landslide hazard would have a low priority in planning areas to be set aside for national parks or game preserves. Conversely, landslides can be an important factor in the development of newly cleared forest areas or in building infrastructure in mountainous or hilly terrain. Clearly, the amount of landslide hazard information needed is based on the level and type of anticipated development for an area. Failure to understand the potential effects landsliding can have on a project or how the project might affect landslide potential can bring increased risk.

Natural changes as well as human-induced changes can affect the susceptibility to landslides and must be understood when assessing the landslide potential of an area. It is critical for a planner to appreciate these issues early in the planning process. A decision is ultimately made regarding the degree of risk that is acceptable or unacceptable to a project. Mitigation strategies are then designed to reduce risk. These concepts are discussed at more length later in this chapter.

Early consultation with landslide technical specialists is recommended so that they can assess the risk of proposed activities in a landslide hazard area. The planner, while not expected to be a technical expert, must know the questions to ask of a landslide specialist. By asking the appropriate questions, the planner will be able to identify and evaluate measures to minimize or avoid landslide vulnerability.

Landslide Hazard Mapping

Interpretation of future landslide occurrence requires an understanding of conditions and processes controlling landslides in the study area. Three physical factors-past history, slope steepness, and bedrock-are the minimum components necessary to assess landslide hazards. It is also desirable to add a hydrologic factor to reflect the important role which ground water often plays in the occurrence of landslides. An indication of this factor is usually obtained indirectly by looking at vegetation, slope orientation, or precipitation zones. All of these factors are capable of being mapped. Specific combinations of these factors are associated with differing degrees of landslide hazards. The identification of the extension of these combinations over the area being assessed results in a landslide hazard map. The technique used to prepare hazard maps is called a combined factor analysis and is described in detail in Section C of this chapter.

Integrating Landslide Hazard Zonation Maps into the Development Planning Process

Landslide hazard information serves as one of the many components in an integrated development planning study. Since landslide activity can adversely affect or interfere with human activity, landslide hazards constrain or limit land-use capability. For this reason, it is important to identify relative landslide hazard levels early in the planning process. This permits planners to determine the degree of landslide risk that is acceptable or unacceptable to a development programme. Decisions can then be made regarding which of these measures will be taken: avoidance, prevention, or mitigation of existing and future landslide hazards in the development programme. The method described in this chapter places emphasis on landslide hazard identification and its use in an integrated planning study as natural resources are evaluated, a development strategy is formulated, and investment projects are identified at the profile level.

Preliminary Mission

During the Preliminary Mission of an integrated development planning study, an initial review is made of the type and content of available information, including natural hazard information. The availability of geologic, topographic, hydrologic, and vegetation maps and aerial photographsis usually ascertained. This information is essential for executing a landslide hazard zonation. Also during this stage of the study, available information should be collected and reviewed concerning assessments of natural hazards, including landslides and disasters, which are known to have affected the study area.

Phase I - Development Diagnosis

In the context of planning the development of a river basin, province, or other planning unit, a development diagnosis assists in identifying areas with highest development potential. These are designated "target areas," in which subsequent, more detailed studies are concentrated. Part of the development diagnosis process involves identifying and delineating natural resource factors that favour or constrain development in a particular area. Landslide hazard is an undesirable factor, and the greater the hazard, the more it may shape the development potential.

When a potential hazard is present in the study area, the first step is to undertake a brief survey to establish whether landslides have occurred in recent times. Roads, railroads, and river banks are good sites for seeking signs of past landslide occurrence. Discussions with local authorities responsible for public works, forestry, and agricultural activities can prove to be a valuable source of information since they may be familiar with past landslides in an area. However, it is important to bear in mind that new development activities may increase landslide hazards, and the absence of evidence from past landslides does not guarantee that landslides will not pose any problems in the future.

Preliminary Mission (Study Design) Questions Planners Need to Ask:

- Are geologic, topographic, hydrologic, and vegetation maps available? At what scale?
- Are aerial photographs available? At what scale?
- Does the study area have a history of landslides and/or disasters caused by landslide events?
- Is landslide hazard assessment information available?

Key Decisions to be Made at This Stage:

- Is it likely that landslides will affect major and/or significant portions of the study area?
- Will the study have access to landslide hazard assessment information other than what it may produce?

PHASE I: *Development Diagnosis (Strategy And Project Identification) Questions Planners Need To Ask:*- Is sufficient information to prepare a landslide inventory map, an isopleth map of existing landslides, and/or a landslide hazard map using factor analysis available? - How will the assessment be carried out? During what time period? How will the assessment information be integrated into the overall study development strategy and project identification activities?

Key Decisions To Be Made At This Stage:- Is a landslide hazard map necessary? - At what scale should the map be prepared? - Who will execute the assessment? - Who will be responsible for incorporating the assessment information into the overall study activities?

The areal extent and variety of development activities being considered make determining the landslide susceptibility based on all existing landslides, regardless of type, an appropriate approach (DeGraff, 1982). A simple inventory of past landslides, along with data regarding bedrock, slope steepness, and - when available-the hydrologic factor, produces a landslide hazard map that will satisfy the needs of the development diagnosis. Suitable scales for the landslide hazard map range from 1:250,000 to 1:50,000.

Having limited or insufficient data for preparing the combined factor analysis is most likely to be a problem encountered at the development diagnosis level. When this situation arises, there are two options: (1) invest the money and human resources to obtain the data needed to produce a landslide hazard map, or (2) prepare an isopleth map of existing landslides (described in Section C of this chapter). The isopleth map shows areas of frequent or infrequent landslide occurrence. While this type of map provides some idea of where landslides can be a major influence on development, it is only a rough approximation of where a problem can be encountered during development. Isopleth maps are an acceptable option at this stage of development but are wholly unsuitable for use in the more detailed planning stages.

The degree of landslide hazard in an area is a limiting factor only for those activities that may alter the existing balance between forces driving and resisting movement on an unfailed slope. Planners need to understand what effects development activities may have on this balance of forces. For example, placing a fence around a field is not going to produce a landslide, nor will it prevent one. Removing forest cover to create a field for cultivating crops is much more likely to lead to landslide occurrence, since it alters the balance of forces and may increase the susceptibility to slope failure by some triggering event, such as prolonged rainfall, which would not have produced the landslide under the original conditions. This increased susceptibility may not be immediately apparent since there may be a lag time before this is evident.

Landslide hazard zonation can be represented as an individual factor limiting land capability or it can be combined with hazard zonation for other natural hazards as an aggregate hazard. There are at least 10 different approaches that have been used to generate land capability maps (Hopkins, 1977).

Planning Stage	***Hazard Identification Need***	***Landslide Inventory Level***	***Suitable Scales For Hazard Maps***
Preliminary Mission	Identify hazard issue	As available	As available
Phase I- Development Diagnosis	Degree of hazard from all types of landslides	Simple	1:250,000 to 1:62,500
Phase II - Action Plan and Project Formulation	Degree of hazard from all types of landslides supplemented by hazard from some specific types	Intermediate	1:62,500 to 1:10,000
Project Implementation	Site-specific hazard based on geotechnical models	Detailed	1:12,500 to 1:500

Chapter 3 discusses land capability in more detail. The method for landslide hazard assessment presented in this chapter results in the production of a map. Thus, it can be considered in the application of land-use capability approaches.

There are two main applications of a landslide hazard assessment in land-use capability that includes relative surveys. First, it is used in overall development planning to emphasize the subjective nature of assigning land-use capability. For example, at the development diagnosis stage, the relative classification of "highest" capability can be assessed in relation to the constraints that possible increased landslide hazards may pose to proposed development activities. Second, it can show where existing development may face some risk previously unidentified. This enables prioritizing mitigation activities to be assigned to different development activities.

Phase II - Development Strategy and Project Formulation

An action plan is defined with the objective of facilitating development of target areas identified in Phase I. Development projects considered for the target area are formulated at this stage. Also at this time, landslide hazard evaluation within the study area is refined. The general landslide hazard assessment must be supplemented with an intermediate inventory to show the degree of hazard for specific types of landslides that may impact on proposed development activities. For example, introducing widespread agricultural activities into a forested environment requires a greater understanding of the hazard from shallow landsliding rather than from deep-seated rockslides.

In developing areas with landslide hazards, mitigation measures should be selected if they are not already part of the project identification information. It is possible to reduce the possible impact of natural landslide activity and limit landslides which occur as a result of human activity (Kockelman, 1985). There are two basic approaches: first, to avoid landslide-susceptible areas, and second, to design measures to compensate for the inducement of landslides. For example, make location decisions so as to avoid building in certain areas, such as placing dwellings and critical infrastructure outside areas with a high likelihood of natural landslide activity. In some instances, the potential effects of a landslide can be mitigated. Landslide hazards resulting from development can be reduced by designing changes to counteract the impact that development may have on slope integrity. This might take the form of permitting only warehouses and storage facilities in higher hazard areas, to reduce the vulnerability to the population should a landslide occur.

Landslide Triggering Actions, Mitigation Approaches, and Development Variables

- Triggering Actions:

- From other hazards:
- Earthquakes
- Floods

- Fire (and resulting loss of vegetation)
- Volcanoes
- Development related:
- Changes in vegetation cover
- Earthen dams
- Excavation and mining
- Irrigation
- Infrastructure and structure construction
- Liquid disposal (sanitary, sewers, latrines, etc.)
- Pilings
- Soils deposits

- Mitigation Approaches:
  Avoidance
- Insurance and taxation
- Land-use zoning
- Stabilization
- Structural design

- Development Variables:
- Available information
- Economic, social, and political concerns
- Existing development
- Proposed development

In formulating investment projects, a more detailed hazard zonation map is needed. An intermediate landslide inventory is needed which provides greater detail to distinguish different landslide types. This data can be used for a reanalysis of the combined factor analysis. This reanalysis yields an improved landslide hazard map. If the hydrologic factor was not part of the earlier landslide hazard analysis, its inclusion at this stage would greatly improve the resulting hazard map.

At this stage, the value of a landslide hazard map to planners can be enhanced by representing areas where specific landslide types are prevalent. This is accomplished by preparing an isopleth map, as mentioned in Phase 1. Preparation, however, should be altered to meet the specific needs of this planning stage. Alteration of the isopleth map is described in detail in Section C's discussion of

"Compensating for Insufficient Data-The Isopleth Map". This produces a map representing the intensity of past landslide occurrence in a form resembling a topographic map. The isopleth lines appear similar to the contour lines showing elevation. The final isopleth map is used as an overlay on the landslide hazard map.

It should be noted that the isopleth map does not alter the basic hazard zones determined previously. It is still an analytic map, which in this instance shows the varying prevalence of a specific landslide type in an area. It provides an additional criterion for the planner to make use of in deciding which area may be best suited for certain development activities. This is especially helpful in evaluating moderate hazard zones.

Phase II: Project Formulation and Action Plan Definition Questions Planners Need to Ask

- Does the initial combined factor analysis landslide hazard map provide sufficient information to proceed with investment project formulation?
- If not already included, is there a hydrologic factor which could be added for further detail on hazard zones?
- Should an isopleth overlay be added to the hazard zonation map?
- Are there certain proposed land uses for which recommendations for mitigation should be included in the formulation of investment projects?

Key Decisions to be Made at This Stage

- Who will execute the intermediate landslide hazard assessment?
- Which areas should be included in the additional assessment?
- Who will be responsible for incorporating additional information in investment project formulation activities

Where proposed land use is recognized as susceptible to a specific landslide type, the proposed activity is best located in a low hazard zone or moderate hazard zone with the lowest occurrence, i.e., smallest isopleth value, of that landslide type. The improved landslide hazard map and isopleth overlay require that an intermediate landslide inventory be prepared at this planning level. The landslide hazard map suitable for formulating development projects should be at a scale of 1:62,500 to 1:12,500.

Project Implementation

The landslide hazard map can contribute to planning for a project's implementation. There are two situations when this map may prove beneficial, both of which are related to mitigating the potential effects of landslides. In one case, it is conceivable that if areas identified with a moderate landslide hazard are targeted for development, greater detail of those areas is needed to ensure the project design compensates for this greater hazard potential. For example, moderate or higher hazard areas may not be entirely avoidable along a proposed road. Detailed investigation can provide information on groundwater conditions and on the stability characteristics of soil and rock to ensure a stable design (Morgenstern and Sangrey, 1978).

In another case, existing infrastructure or communities may be located in previously unidentified high hazard zones. These areas should be given priority for introducing some measure of mitigation.

For example, the effect of landslides issuing into an inhabited area from nearby mountain canyons might be mitigated by constructing debris basins to trap most of the material. Where such mitigation is impossible and the risk is identified as being extremely high, relocation to a safer area may be considered.

A detailed hazard map for the specific site in question is necessary at this stage of project design. Preparation of a detailed landslide inventory is now necessary. The large-scale features represented on landslides mapped in this detailed inventory are valuable for test drilling of a site and other sampling activities of engineering design work. Detailed landslide inventories and related interpretation of test results require map scales from 1:12,500 to 1:500.

The next section provides a detailed discussion of the types and nature of landslides, the basis for assessing landslide hazards, and the factors associated with landslide activity.

Landslides, Landslide Hazard Assessment, and Areas of Concern

Landslides and Landslide Susceptibility: Landslides are caused when the force of gravity pulls rock, debris or soil down a slope. They are one of the forms of erosion called mass wasting, which is broadly defined as erosion involving gravity as the agent causing movement. Because gravity constantly acts on a slope, landslides only occur when the stress produced by the force of the gravity exceeds the resistance of the material. This is distinct from some other forms of erosion caused by running water, for instance, which occurs when precipitation falls on a slope or within a channel carrying a stream or river.

Project Implementation Questions Planners Need to Ask:

- What type of landslide problems exist?
- What site-specific conditions need to be known for final design of an investment project with low landslide vulnerability?

Key Decisions to be Made at This Stage:

- Who will execute the detailed landslide hazard assessment?
- Which mitigation measures should be considered to reduce the risk to an acceptable level?
- Who will be responsible for incorporating the additional information in project implementation activities?

Landslide movement is perceptible and may take the form of falls, topples, slides, or flows. It can consist of free-falling material from cliffs, broken or unbroken masses sliding down mountains or hillsides, or fluid flows. Materials can move up to 120mph or more, and slides can last a few seconds or a few minutes, or can be gradual, slower movements over several hours or days. Accordingly, landslides are recognized on the basis of type of movement. The most widely used classification scheme divides landslides into different types according to the material being moved and type of movement (Varnes, 1978). Speed of movement and amount of water mixed with the material are secondary parametres defining some landslide types. Recognizing the types of landslides presents in an area helps explain how and where factors have contributed to natural slope instability in the past. Factors influencing where landslides occur can be divided into two sets, permanent and variable (Sharpe, 1938). Permanent factors are characteristics of the landscape which remain unchanged or vary little from a human perspective. The steepness of a slope or the type of rock, for example, presents changes only with the passage of long periods of time. Permanent factors such as rock type and slope steepness can be recognized and identified for specific landslides long after their occurrence (DeGraff, 1978). By examining existing landslides in an area, it is possible to recognize how permanent factors contributed to these slope failures. Identifying conditions and processes promoting past instability makes it possible to use these factors to estimate future landslides (Varnes, 1985). Variable factors are landscape characteristics that change quickly as a result of some triggering event. Ground vibration due to earthquakes, a rapid rise in groundwater levels, and increased soil moisture due to intense precipitation are examples of variable factors. It is often necessary to be present at the time a landslide occurs or shortly thereafter to assess these factors. Due to the lack of long-term records relating landslide activity to historic earthquakes, storms, or other initiating factors, permanent factors are usually used to estimate landslide hazard. As such, identifying landslide areas is not an accurate science and leads, in general, to depicting hazard-prone areas based on an estimation. At best, landslide and landslide susceptible areas can be identified along with expected triggering events. At worst, some areas may not be detected at all.2. Hazard Assessment of Landslides Landslides are not currently amenable to risk assessment since there is no basis to determine the probability of landslides occurring within a given time period. Hazard assessments are possible and can be used in place of risk assessments. Hazard assessments are estimations of an area's susceptibility to landslides based on a few key factors. These are each capable of being mapped and allow land areas to be evaluated on their relative susceptibility to landslides. Figure given,

Definition Of Basic Landslide Termsnomenclature: A steep surface on the undisturbed ground around the periphery of the slide, caused by the movement of slide material away from undisturbed ground. The projection of the scarp surface under the displaced material becomes the surface of rupture. Minor Scarp: A steep surface on the displaced material produced by differential movements within the sliding mass. Three principles guide landslide hazard assessment. First, landslides in the future will most likely occur under geomorphic, geologic, and topographic conditions that have produced past and present landslides. Second, the underlying conditions and processes which cause landslides are understood. Third, the relative importance of conditions and processes contributing to landslide occurrence can be determined and each assigned some measure reflecting its contribution (Varnes, 1985). The number of conditions present in an area can then be factored together to represent the degree of potential hazard present.

Landslide hazard has been determined with a high degree of reliability for only a few locations. These have required careful, detailed study of the Interaction of pertinent permanent and variable conditions in the target area. This can be a very expensive and time-consuming process that is unjustified for the purpose of broad-scale development planning. Landslide hazard zonation is one technique that can be used in the early stage of a planning study.

Most assessment procedures for landslide hazard zonation employ a few key or significant physical factors to estimate relative landslide hazard. The method described here requires a minimum of three factors mentioned earlier: distribution of past landslides, type of bedrock, and slope steepness, and a fourth, hydrologic factor may be added to reflect the important role which groundwater often plays in landslide occurrences (Varnes, 1985, and USGS, 1982).

Each factor is represented in a quantitative or semi-quantitative manner to facilitate the identification of different degrees of landslide hazard in an area.

Since all of these are permanent features, it is usually possible to map each factor. Specific combinations of these factors can be associated with differing degrees of landslide hazard. By extending these combinations over an entire area, a landslide hazard map is produced.

Factors Associated with Landslide Activity

The distribution of past landslides within the area, type of bedrock, and slope steepness represent, respectively, geomorphic, geologic, and topographic factors (Varnes, 1985, and USGS, 1982). Each of these factors is described in more detail below to give the planner a better understanding of their contribution to landsliding. The final section, "C. Mapping Physical Factors and Preparation of a Landslide Hazard Map," provides information on mapping them.

Past Landslides and Their Distribution

Interpreting the likelihood of future landslide occurrences requires an understanding of conditions and processes controlling past landslides in the area of interest. This can be achieved by examining and mapping past landslide activity in the area. Geologic, topographic, and hydrologic circumstances associated with past landslides indicate which natural or artificially created circumstances are likely to produce landslides in the future.

A primary consideration of the planner is the effect of existing land use on landslide activity. Certain types of landslides may be associated with specific land uses. For example, certain slides may only occur in road cuts or excavations. There may even be a critical height-to-inclination relationship for cutslopes below which these landslides will not occur. Field studies can provide insight into how different factors have contributed to failures. In some investigations special forms have been employed to ensure consistent collection of this ancillary information. A summary of observations about landslide conditions and processes are incorporated into each landslide inventory, e.g., as in Pomeroy (1979), and mapped.

In Order to Plan for Development in Landslide Areas, a Planner Needs to Know

- Past distribution of landslides
- Types of bed rock
- How steep the slopes are
- Indirect measures of the area's hydrologic characteristics that are available
- Effects development activity can have on landslide susceptibility

Bedrock

Bedrock influences landslide occurrence in several ways. Weak, incompetent rock is more likely to fail than strong, competent rock. On slopes where weak rock overlain by strong rock is exposed, the difference in strength increases the potential for landsliding in the stronger rock as well since the weak rock tends to erode and undermine the stronger rock.

The strength of a rock mass depends on the type of rock and the presence and nature of discontinuities such as joints or other fractures. The more discontinuities present in bedrock, the greater the likelihood of rock instability. Rock type may exert control on landsliding by influencing the strength of surface material in the area. For example, soils (in the engineering rather than agricultural sense of the term) derived from schists or shales will contain high percentages of clay. These soils will have different strength characteristics than coarser-grained soils such as those derived from granitic bedrock. There are many ways, then, that rock type or structure contributes to the instability, which can be represented on a map.

Slope Steepness or Inclination

The influence of slope steepness on landslide occurrence is the easiest factor to understand. Generally, steeper slopes have a greater chance of landsliding. This does not prevent failures from occurring on gentler slopes. Other factors may make a gentle slope especially sensitive to failure, and thus in this situation could be determined to have a relatively high hazard potential.

For example, high ground water conditions occurring in sandy soils may liquefy during an earthquake. This can cause a landslide on a slope as gentle as 5 to 10 percent. Conversely, the steepest slopes may not always be the most hazardous. Steep slopes are less likely to develop a thick cover of superficial material conducive to certain types of landslides. Slope steepness can be mapped using generally available topographic maps.

Hydrologic Factor

Water is recognized as an important factor in slope stability-almost as important as gravity. Information on water table levels and fluctuations is rarely available. To represent the hydrologic factor in landslide hazard assessments, indirect measures can be used which can be mapped to show the influence of the area's hydrology, such as vegetation, slope orientation (aspect), or precipitation zones.

The type of vegetation and its density over an area will often reflect the variation in subsurface water. Certain species are water-loving or phreatophytes. Presence of these species shows near-surface water table conditions and springs. In mountainous regions, microclimatic differences produce different hydrologic conditions which in turn result in plant communities that vary according to the moisture available to the slope and its distribution throughout the year.

Slope orientation (aspect) refers to the direction a slope faces. It can be an indirect measure of climatic influence on the hydrologic characteristics of the landscape. Important characteristics associated with landslides are related to such factors as subsurface recharge resulting from prevailing winds and their influence on local frontal storms or accumulated snow. In other cases, a slope may experience more freeze/thaw cycles or wet/dry cycles which can reduce the strength of the soil and make the area more susceptible to landslides. In general, due to the complexity of these factors and existing development activities, there is usually no direct observable correlation between slope orientation and landslide hazard.

Human-Initiated Effects

In addition to natural phenomena, human activities may increase the natural tendency for a landslide to occur. Landslides which result from development activities are usually the result of increasing moisture in the soil or changing the form of a slope. Development activities such as cutting and filling along roads and the removing of forest vegetation are capable of greatly altering slope form and ground water conditions (Swanson and Dyrness, 1975). These altered conditions may significantly increase the degree of landslide hazard present (Varnes, 1985, and Sidle, Pearce, and O'Loughlin, 1985).

For example, converting a forested area to grassland or one where crops are cultivated can increase the moisture in the soil enough to cause landslide problems (DeGraff, 1979). Or building a road which cuts off the toe of a steep slope can increase landslide susceptibility. It is possible to reduce the potential impact of natural landslide activity and limit development-initiated landslide occurrence by early consideration of these effects (Kockelman, 1985).

Now that the general points with regard to mapping the various land characteristics have been covered, the final section provides details on the techniques to do so in addition to presenting a step-by-step approach for preparing a landslide hazard map.

Mapping Physical Factors and Preparation of a Landslide Hazard Map

A landslide inventory produces a descriptive or data map (Cotecchia, 1978). By overlaying the landslide inventory map on the maps of the type of bedrock, slope steepness, and indirect hydrologic measures, the association of past landslides with the factors controlling landslide occurrence can be recognized. The method described below employs these associations in synthesizing a landslide hazard map. Extrapolating the data to areas with characteristics similar to those found associated with past landslides is an effective tool for forecasting where, but not when, landslides are more likely to occur in the future.

This section presents the techniques used to map each of the key factors associated with landslides. With these maps, a landslide hazard map can be prepared. Hazard zonation is a means of identifying areas with differing landslide hazards. The step-by-step approach, or factor analysis, used to prepare a landslide hazard map is described.

Mapping the Physical Factors Associated with Landslides

Each factor is mapped separately by a different technique.

Mapping the Inventory of Existing Landslides: A map of existing landslides serves as the basic data source for understanding conditions contributing to landslide occurrence. Normally such a map is prepared by the interpretation of aerial photography and field examination of selected locations. While this map could also be compiled by field methods alone, the time and expense involved would only be justified by the unavailability of photo coverage. Either means of map preparation requires the skills of a geologist with experience in landslide or landform interpretation.

Aerial photography can serve as the source for data on existing landslides, type of bedrock, and vegetation cover. Typically, large-scale photography is necessary to be useful for existing landslides. The photo scale depends on the size of landslides common to the study area. Small-scale photography is less of a concern where bedrock and vegetation exist, since delineating areas with similar texture and appearance is easier than recognizing discrete features. Satellite imagery is generally unsuitable for landslide mapping except where data products can be enlarged to at least 1:50,000 scale. Photographic and satellite information is valuable in mapping other spatial information and for use in conjunction with computer mapping techniques as part of the development planning study.

Depending on vegetative cover, photo quality, and the skill of the interpreter, overall identification accuracy of 80 to 85 percent is realistic using aerial photography (Rib and Liang, 1978). The range of useful scales of aerial photography for landslide inventory work is limited to about 1:40,000 or larger. The selected scale will depend on the size of landslides common to the study area and, to some extent, the relief of the area. Large failures of four or more square kilometres are extremely difficult to detect on aerial photography smaller than 1:40,000. Where the majority of landslides are one hectare or smaller in size, large-scale photography on the order of 1:4,800 is necessary. The usefulness of black and white, colour, or colour-infrared photography for landslide inventory work will vary with local conditions and the individual making the interpretation. Each type of photography has advantages and disadvantages that will vary in their importance according to the characteristics of the area being mapped.

The map may be prepared at different levels of detail concerning existing landslides (USGS, 1982). A simple inventory identifies the definite and probable areas of existing landslides and is the minimum level required for a landslide hazard assessment. A map is produced in which each landslide is outlined and an arrow is drawn to denote the direction it moved.

More information can be provided by producing an intermediate inventory. The map produced at this level would show the outlined landslide types and distinguish between areas of landslide origin and deposit. The former is the area where material once existed as the source of the landslide and appears as a scar. The latter is deposited material from the landslide. The most information is obtained by producing a detailed inventory (Wieczorek, 1984). Large-scale features such as secondary scarps, sag ponds, and ground-crack patterns may be represented on individual landslides.

These three types of inventories can be prepared as the development study progresses. To reiterate what was presented in Section A of this chapter: the simple inventory is adequate for Phase I development diagnosis activities; the intermediate inventory provides greater detail for an improved hazard map of a target area in Phase II; and the large-scale features of the detailed inventory are necessary for final project design in the implementation stage.

There are several considerations to keep in mind when gathering data on existing landslides. First, the time and effort required to conduct an inventory varies with (1) geologic and topographic complexity; (2) size of an area; and (3) desired level of inventory detail (Varnes, 1985). Figure depicted, characterizes the relationship between the amount of time and level of effort for these three variables. Second, more detailed inventories will require larger map scales to reveal the small features of this added detail. Third, additional data gathering can add detail to an existing inventory. This enables a previously completed simple inventory to be transformed into an intermediate inventory with less time and effort than producing the intermediate inventory solely from field work and aerial photography.

Mapping the Types of Bedrock Contributing to Instability

By using bedrock as one factor in the landslide hazard assessment, the many different ways rock type or structure contribute to instability are represented. Comparing a bedrock map with the landslide map, one can discriminate between rock units associated with existing landslides and those devoid or largely free of landslide activity.

To produce a usable bedrock map for a hazard assessment, bedrock unit boundaries should be traced to produce new, more suitable units. Existing standard geologic maps define units according to factors such as age, composition or lithology (rock type), and structure (faulting, folding, etc.). For example, a standard geologic map may show a series of volcanic ash deposits of similar mineral compositions which vary only slightly in age. In most instances, these different units will affect landslide occurrence in a similar way and should be delineated as a single bedrock unit in a revised map for hazard assessment work. The geologist must use professional judgement to ensure that the number of bedrock units is sufficient to distinguish differences in their effect on landslide occurrence.

When a geologic map does not exist, a bedrock map based on aerial photography with limited field verification is needed. This map may be no more detailed than a delineation of sedimentary, igneous, and metamorphic rock types. Obviously, a bedrock map generalized from a more detailed map is preferable, but this is an acceptable alternative under such circumstances. Delineating areas with similar texture and appearance is easier than recognizing discrete features. Scales as small as 1:62,500 are useful for this work. Photos at scales of 1:20,000 or larger are difficult to use because the limited area shown restricts comparison with adjacent contrasting areas. It also significantly increases the number of photographs to examine in mapping the area. Black and white, colour, and colour infrared photography are all suitable for bedrock mapping. Satellite imagery is generally unsuitable for this mapping except when the imagery is enlarged to usable scales. For example, imagery at a 1:50,000 scale produced from satellite imagery is acceptable for this mapping.

A soils map is an inadequate substitute for a bedrock map. Soils maps are based on factors concentrated in the upper metre or less of superficial material that affects agricultural activities. Generally, there is little or no correlation between "agricultural" soil characteristics and the likelihood of failures originating along surfaces a few metres to tens of metres deep in superficial material.

Scale 1:50.000

EXPLANATION

Landslide

Figure: *Simple Landslide Inventory Map*

Mapping Slope Steepness or Inclination

Slope steepness is a factor which associates the effectiveness of gravity acting on a slope to landslide susceptibility. A topographic map is the source for preparing a slope steepness map. The slope steepness map displays the steepness values associated with the majority of existing landslides and is derived from an existing topographic map. Steepness for landslide hazard assessments is commonly expressed as a percentage rather than in degrees. The categories or grouping of steepness values for use in analyzing landslide hazards should approximate those of the slopes present in the study area. Too many classes will make it difficult to identify slopes critical to landslide occurrence and too few will be equally useless.

The Optional Hydrologic Factor-Mapping Indirect Measures

Since information on water table levels and fluctuations is rarely available, mapping indirect measures such as vegetation and slope orientation can reveal the influence of hydrology on an area. Any vegetation map used to represent the hydrologic factor in the landslide hazard assessment must employ units that are dependent on water. This may be as simple as representing phyreatic and non-phyreatic plant communities or as complex as distinguishing different forest types. Selection of the appropriate vegetation map units to indicate the effects of water in causing landslides requires careful field observation by the geologist.

Aerial photography is an appropriate source of data for preparing vegetation maps. In preparing vegetation maps, as in mapping bedrock, scale is less of a concern. Here, too, delineating areas with similar features is easier than recognizing discrete features. Scales of 1:62,500 are useful in identifying vegetation since 1:20,000 and larger scales do not reveal the contrasting characteristics of adjacent areas. Black and white, colour, and colour-infrared photography are all suitable for this mapping. Satellite imagery is acceptable only when enlarged to usable scale.

The direction in which a slope faces can also be -mapped and used as an indirect indicator of the hydrologic factor. Slope orientation or aspect is described in terms of the eight cardinal directions, i.e., north, northeast, etc. For convenience in establishing a data base, slope orientation is measured in degrees of azimuth from 0 to 360 degrees. Each cardinal direction is defined by a set of azimuth values. For example, slopes facing the northeast can have an azimuth reading ranging from 22.5 degrees to 67.5 degrees.

Interpreting Landslide Hazards: The Landslide Hazard Map

A landslide hazard map is generated to identify areas with differing landslide hazards. A hazard map is produced for each stage of the planning process, from the more generalized map in the initial stage to a detailed zonation map for specific site use. As the name suggests, this map divides the entire study area into sub-areas based on the degree of a potential hazard from landslides. The landslide hazard map is produced by interpreting the data represented by the maps of the inventoried landslides and the permanent factors found to influence the occurrence of landslides. As with any map, scale is an important consideration. There are two points to keep in mind concerning the scale of the landslide hazard map. First, such a map should be produced at a scale capable of representing the information needed at a particular planning level. Compatibility of scale would be important when the hazard map is to be combined with other maps to yield a land capability map. Second, the landslide hazard map should be at a scale not markedly different from the data maps used to produce it. In other words, reliability may be questionable when a landslide hazard map produced at a scale of 1:50,000 has been based on a 1:250,000 slope steepness map.

Four levels of relative hazard are identified on a landslide hazard map: (1) low; (2) moderate; (3) high; and (4) extreme hazard. The level of landslide hazard is measured on the ordinal scale with this method and is a quantitative representation of differing hazard levels that shows only the order of relative hazard at a particular site and not absolute hazard. Predicting absolute hazard is impractical with current capabilities.

As a consequence, there is no way to compare hazard zones at different sites or to determine the likelihood that a high hazard area, for example, is two or ten times more likely to fail in the future than low hazard areas. It should be stressed that these relative hazard zones are based on the existing landslides and conditions influencing their occurrence in a specific area. The hazard zones which are determined for an area hold true only for the area for which they were prepared. Similar conditions found outside the assessed area may not produce the same degree of hazard because of some seemingly minor difference in one of the factors.

Factor Analysis: The Technique to Prepare a Hazard Map

A factor analysis is a step-by-step approach used to prepare a landslide hazard zonation map of an area.

There are four steps to complete the factor analysis and produce a hazard map: (1) map the existing landslides and prepare a map combining the permanent factors (bedrock, slope steepness, and, when available, the hydrologic factors) into individual map units; (2) overlay the landslide inventory on the combined factor map; (3) prepare a combined factor analysis for all combinations of the factors and group combinations of these factors in a way that defines the four levels of landslide hazard; and (4) produce a map with four landslide hazard zones from the grouped combinations.

Step One: Combined Map of Permanent Factors

The first step is to prepare a map of the inventoried existing landslides. Also, compile a map which combines the bedrock, slope steepness, and, when included, hydrologic factor units or categories into individual cartographic units. As an example, assume that only bedrock and slope steepness are being used. The compiled map will be composed of cartographic units delineating certain bedrock type and slope values, e.g., Bedrock B3 on slopes between 25-50 percent.

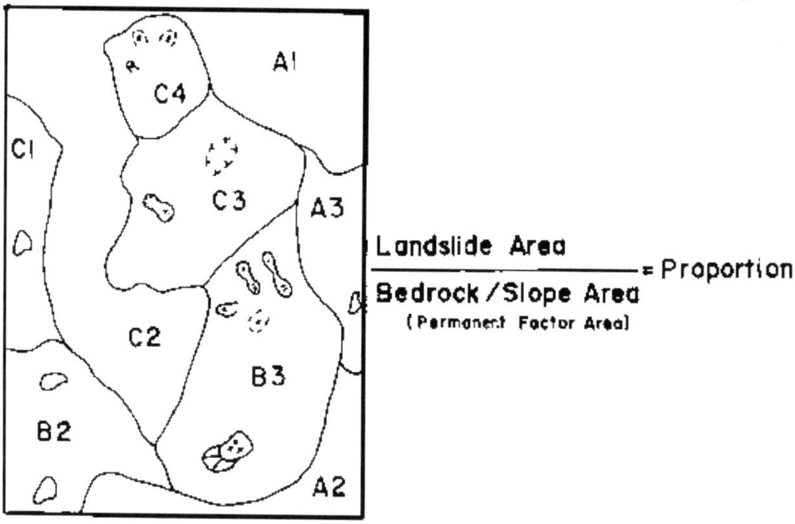

Figure: *Study Area Map*

Legend

A representation of how the proportion of bedrock slope combinations subject to past landslide activity is determined. Note that while combination B3 obviously has more landslides than combination C4, the smaller size of C4 area will likely result in its having a higher proportion than B3.

Step Two: Overlay of Landslide Inventory

The second step is to overlay the landslide inventory map with this combined factor map. This will identify which combinations are associated with past landslides and which are not. A landslide inventory table is developed indicating total area of landslides occurring on each specific bedrock unit and slope steepness combination (and other factors, if considered). When a hydrologic factor such as vegetative zone or slope orientation is used, the table will include the area of landslides for each specific combinations of bedrock, slope steepness, and the hydrologic factor. Summing the areas from all combinations found in the table will yield the total area of landslides in the study area. This is a way to check that all combinations are included in the analysis. Figure shown earlier, shows the extent to which each combination is present in the study area. For example, Bedrock B on slopes between 25 and 50 percent has 784 hectares of landslides.

Step Three: Group Combinations Using a Factor Analysis

The third step is to group combinations of these factors in a way that will define four levels of landslide hazard. This grouping is achieved by performing a combined factor analysis or matrix assessment (DeGraff and Romesburg, 1980). This analysis permits incorporating the interaction among factors affecting landslide occurrence without explicitly understanding those interactions.

To start, measure the total area for every combination of bedrock, slope steepness, and hydrologic factors in the study area represented in the table prepared in Step 2. The total area with these combinations is to be calculated, not just those associated with landslide activity. Continuing with the example, assume a total area of 2,327 hectares of bedrock B on slopes greater than 25 percent but less than 50 percent was found. The landslide inventory table prepared in Step 2 shows only the area of past landslides present for each combination. Then the total area for every combination associated with landslides found in the landslide inventory table is divided by the area for the same combination of factors found in the study area. In the example, this would be 784 divided by 2,327. This yields a proportion of each combination which is subject to past landslide occurrence, e.g., 0.34. This represents the proportion of the combination disturbed by past landslides in that area.

The combination of bedrock, slope steepness, and hydrologic factors associated with the largest area disturbed by landslides may not, in fact, be the most hazardous: it may simply be the combination which is most common to the study area. Since such an area is the most prevalent combination, it has the greatest chance for being associated with past landslides rather than being most hazardous. The process described above ensures that comparison of landslide hazard among different combinations takes place on an equal basis.

Table: *Combined Permanent Factor (Sample Bedrock And Slope Class) And Land Area Coverage (In Hectares)*

	SLOPE CLASS				
BEDROCK GROUP	0<12% (1)	12<25% (2)	25<50% (3)	>50X (4)	TOTAL AREA (HA)
A	-	52	78	-	130 Landslide Area
	1,570	722	512	237	3,041 Combined Area [a/]
B	-	301	784	-	1,085 Landslide Area
	-	1,776	2,327	-	4,103 Combined Area [a/]
C	78		351	180	609 Landslide Area
	673	2,450	1,790	793	5,706 Combined Area [a/]

[a/] Combined Area = Combined Permanent Factor Area

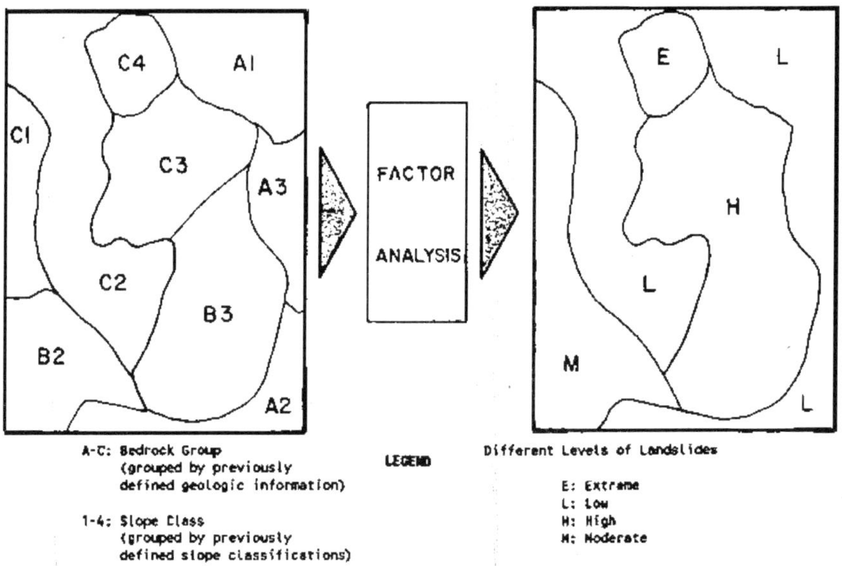

Figure: *Landslide Hazard Zones*

There will be a proportional value for each combination of bedrock, slope steepness, and other factors associated with existing landslides ranging from 0.1. to 1.0. The proportions are sorted from the smallest to the largest. This range of values is broken into three groups to represent the relative landslide hazard in the study area. To ensure that the points used to define the three groups are determined objectively, a non-hierarchical cluster analysis is used.

An initial division into three groups is achieved by breaking equally the range of proportional values present. The upper and lower boundaries of each group are retained or adjusted to ensure that the final division represents the minimum sum of the squared deviations around the three group means. This is based on the W function (Anderberg, 1973).

Step Four: Producing Landslide Hazard Zones

The fourth, and final, step uses the grouped combinations to produce landslide hazard zones-extreme, high, moderate, and low. Once the proportions are divided into three groups, bedrock, slope steepness, and hydrologic factor combinations representing different degrees of relative landslide hazard are identified. The group of proportions with the larger values, i.e., towards the 1.0 end of the range, represent combinations defining extreme landslide hazard. The group of proportions with the next smaller values represents combinations defining high landslide hazard. The group of proportions with the smallest values, i.e., towards the 0.1 end of the range, represents combinations defining moderate landslide hazard. All bedrock, slope, and hydrologic factors not found to be associated at all with existing landslides define low landslide hazard.

The W Function to Define Four Levels of Landslide Hazard

$$W = \sum_{i=1}^{i=3}\sum_{j=1}^{j=n_i} (X_{i,j} - X_i)^2 = W_1 + W_2 + W_3$$

The map overlays used to determine areas of bedrock, slope steepness, and hydrologic factor present in the entire study area can now be revised to make the hazard zonation map. Figure depicted ealier, shows the original maps redrawn into hazard zones. Combinations with extreme hazard are redrawn and relabeled as extreme hazard zones. Redrawing and relabeling for combinations representing other hazard zones produces a completed hazard zonation map displaying four levels of relative hazard. The empirical relationship of the physical factors, as defined by the factor analysis, is valid for only the area evaluated, and cannot be extrapolated to cover additional areas.

Once these hazard areas are identified, a decision can be made regarding the appropriate development activities, type of mitigation measures to be included in the process, or the areas which should be avoided. It is important to note that the essential bedrock and slope steepness maps are not always available. Without these maps, an isopleth map can be produced which is an acceptable substitute.

Compensating for Insufficient Data: The Isopleth Map

In the absence of bedrock and slope steepness maps, the landslide inventory map can be used to produce an analytical map suitable for representing landslide activity in an area. An isopleth map of landslide frequency is recommended for this purpose. An isopleth or any other analytic map can only serve as an initial assessment of landslide activity and not as a substitute for a landslide hazard map. The underlying conditions producing landslides will remain unknown and prevent making the distinction between the relative degrees of landslide hazard.

It is reasonable to assume that areas with a high frequency of landslide activity represent areas with a greater chance of future landslides than those areas with a low frequency. An isopleth map can be made based on this assumption. Preparing an isopleth map begins with the map of inventoried landslides (Wright etal, 1974). A transparent overlay with a 2cm x 2cm grid is placed over the landslide inventory map. At each grid intersection a transparent gridded circle 2.5cm in diametre is centred on each grid intersection on the transparent overlay. The number of grid squares in the circle through which landslide deposits are visible is counted. Divide this number by the total number of grid squares within the inscribed circle. This yields the proportion of the unit area within the circle that is underlain by landslide deposits. This proportion is multiplied by 100 and rounded off to the nearest whole number to compute the percentage of landslide-disturbed terrain. The percent value is written on the gridded overlay next to the grid intersection.

Once all grid intersections are marked with percent values, the isopleth lines can be drawn. Isopleth lines connect the points of equal value. These show the generalized frequency of landslide activity as represented by the percent of landslide-disturbed area. The interval between isopleths drawn to produce the map will depend on the proposed use. A single value representing a boundary between areas of frequent landsliding and infrequent landsliding shows areas where this phenomenon is a major factor in shaping the landscape and areas where it is not. This serves as an initial assessment of areas subject to landslide problems when additional factors are not available for a study area. It is important to remember that this is an analytic technique producing a limited assessment of an area rather than a technique developed by an interpretative process.

During Phase II of the planning process, in addition to the intermediate landslide inventory, the preparation of an isopleth map which would enhance the information available to planners is recommended. Using the technique described above, preparation is altered in two ways: (1) only the specific landslide types identified in the intermediate inventory which are likely to be initiated by the land use proposed would be used in compiling this isopleth map; the choice of landslide types should be governed by the information on landslide activity developed by the geologist completing the intermediate inventory of existing landslides and by existing and proposed land use; and (2) isopleths are drawn at regular intervals similar to the way elevation is represented by a contour interval instead of the single value used in the isopleth map. For example, an interval of 10 percent has been used with some isopleth maps applied to land-use planning (Campbell, 1980, and Pomeroy, 1978). This produces a map representing the intensity of past landslide occurrence in a form resembling a topographic map. The isopleth lines would appear like the contour line showing elevation. The final isopleth map is used as an overlay on the landslide hazard map.

Computer-Generated Mapping

The method described in this chapter can be readily adapted to computer-generated mapping (Brabb, 1984). The factor maps used to generate the landslide hazard map can be encoded to a geographic information system (GIS) and manipulated by a computer. This enables the rapid preparation of tables showing the area for different factor combinations. In some cases, data maps used in landslide hazard assessment may be part of the GIS created for generated land-use planning, for instance a vegetation map. A second advantage of this approach is that the scales for maps to be overlaid in a landslide hazard assessment can be matched regardless of their original scale. For example, the scale of a published bedrock map may differ from the other factor maps. Using manual techniques, redrafting the bedrock map at the same scale would be necessary, whereas a computer-based system permits the matching of map scales regardless of their original scale and the maps can be overlaid.

Computerized matching of map scales requires that certain reference points on each map be identified to ensure proper registration of points between maps. Once maps are computerized, they are capable of being updated or used to improve landslide hazard assessments.

A more detailed landslide inventory map could be encoded and used to produce an improved hazard zonation map with the already encoded data maps.

The single major limitation of using a computer-based system is the amount of time and expense that is required to encode the maps and establish the data base for a landslide hazard assessment at a scale sufficiently large to permit the calculation of the percentage of the area covered by existing landslides. Creating such a data base usually dictates that a major project or series of projects be planned to justify this commitment of resources, or that a data base of computerized maps already exists. One final consideration is the ability to gain access to computer equipment, since computers may be scarce or in great demand for many uses. Nevertheless, readily available and relatively inexpensive micro-computers and software programmes which are adequate for a landslide hazard assessment make it possible for some planning studies to have their own system.

Areas susceptible to landslides can be projected, based on the physical factors associated with landslide activity: past landslide history, bedrock, slope steepness, and hydrology. Predicting where and when landslides are going to occur is not possible even with the best available information. It is, however, possible to identify landslide-susceptible areas. This chapter has discussed some of the concepts related to landslide susceptibility: the different types of landslides; the relative nature of landslide hazard zonation; its relationship to development activity; and ways to mitigate the effects of landslides. The essential point has been to demonstrate the importance of considering landslides early in the planning study and to provide one technique which can be used at all stages of the planning process. The different questions that need to be asked at the different planning stages were highlighted. Many answers can be generated from the use of landslide hazard zonation at each stage of the planning study. The step-by-step combined factor analysis to prepare hazard maps was presented. All of this will enable the planner to have a working knowledge of terms, concepts, and the important considerations related to landslides and landslide hazard mapping.

Landslides and Water

Slope saturation by water is a primary cause of landslides. This effect can occur in the form of intense rainfall, snowmelt, changes in ground-water levels, and waterlevel changes along coastlines, earth dams, and the banks of lakes, reservoirs, canals, and rivers.

Landsliding and flooding are closely allied because both are related to precipitation, runoff, and the saturation of ground by water. In addition, debris flows and mudflows usually occur in small, steep stream channels and often are mistaken for floods; in fact, these two events often occur simultaneously in the same area.

Landslides can cause flooding by forming landslide dams that block valleys and stream channels, allowing large amounts of water to back up. This causes backwater flooding and, if the dam fails, subsequent downstream flooding. Also, solid landslide debris can "bulk" or add volume and density to otherwise normal streamflow or cause channel blockages and diversions creating flood conditions or localized erosion. Landslides can also cause overtopping of reservoirs and/or reduced capacity of reservoirs to store water.

Landslides and Seismic Activity

Many mountainous areas that are vulnerable to landslides have also experienced at least moderate rates of earthquake occurrence in recorded times. The occurrence of earthquakes in steep landslide-prone areas greatly increases the likelihood that landslides will occur, due to ground shaking alone or shaking- caused dilation of soil materials, which allows rapid infiltration of water. The 1964 Great Alaska Earthquake caused widespread landsliding and other ground failure, which caused most of the monetary loss due to the earthquake. Other areas of the United States, such as California and the Puget Sound region in Washington, have experienced slides, lateral spreading, and other types of ground failure due to moderate to large earthquakes. Widespread rockfalls also are caused by loosening of rocks as a result of ground shaking. Worldwide, landslides caused by earthquakes kill people and damage structures at higher rates than in the United States.

Landslides and Volcanic Activity

Landslides due to volcanic activity are some of the most devastating types. Volcanic lava may melt snow at a rapid rate, causing a deluge of rock, soil, ash, and water that accelerates rapidly on the steep slopes of volcanoes, devastating anything in its path. These volcanic debris flows (also known as lahars) reach great distances, once they leave the flanks of the volcano, and can damage structures in flat areas surrounding the volcanoes. The 1980 eruption of Mount St. Helens, in Washington triggered a massive landslide on the north flank of the volcano, the largest landslide in recorded times.

How to Reduce the Effects of Landslides

Vulnerability to landslide hazards is a function of location, type of human activity, use, and frequency of landslide events. The effects of landslides on people and structures can be lessened by total avoidance of landslide hazard areas or by restricting, prohibiting, or imposing conditions on hazard-zone activity. Local governments can reduce landslide effects through land-use policies and regulations. Individuals can reduce their exposure to hazards by educating themselves on the past hazard history of a site and by making enquiries to planning and engineering departments of local governments. They can also obtain the professional services of an engineering geologist, a geotechnical engineer, or a civil engineer, who can properly evaluate the hazard potential of a site, built or unbuilt.

The hazard from landslides can be reduced by avoiding construction on steep slopes and existing landslides, or by stabilizing the slopes. Stability increases when ground water is prevented from rising in the landslide mass by (1) covering the landslide with an impermeable membrane, (2) directing surface water away from the landslide, (3) draining ground water away from the landslide, and (4) minimizing surface irrigation. Slope stability is also increased when a retaining structure and/ or the weight of a soil/rock berm are placed at the toe of the landslide or when mass is removed from the top of the slope.

Slope Stabili*ty*

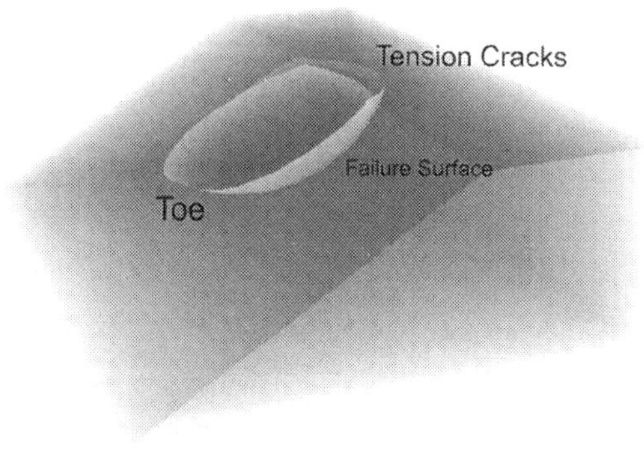

Figure: *Simple slope slip section*

The field of slope stability encompasses the analysis of static and dynamic stability of slopes of earth and rock-fill dams, slopes of other types of embankments, excavated slopes, and natural slopes in soil and soft rock. Slope stability investigation, analysis (including modeling), and design mitigation is typically completed by geologists, engineering geologists, or geotechnical engineers. Geologists and engineering geologists can also use their knowledge of earth process and their ability to interpret surface geomorphology to determine relative slope stability based simply on site observations.

Earthen slopes can develop a cut-spherical weakness area. The probability of this happening can be calculated in advance using a simple 2-D circular analysis package. A primary difficulty with analysis is locating the most-probable slip plane for any given situation. Many landslides have only been analyzed after the fact. More recently slope stability radar technology has been employed, particularly in the mining industry, to gather real time data and assist in pro-actively determining the likelihood of slope failure.

Real Life Failures

Figure: *Real life landslide on a slope*

Real life failures in naturally deposited mixed soils are not necessarily circular, but prior to computers, it was far easier to analyse such a simplified geometry. Nevertheless, failures in 'pure' clay can be quite close to circular. Such slips often occur after a period of heavy rain, when the pore water pressure at the slip surface increases, reducing the effective normal stress and thus diminishing the restraining friction along the slip line. This is combined with increased soil weight due to the added groundwater. A 'shrinkage' crack (formed during prior dry weather) at the top of the slip may also fill with rain water, pushing the slip forward. At the other extreme, slab-shaped slips on hillsides can remove a layer of soil from the top of the underlying bedrock. Again, this is usually initiated by heavy rain, sometimes combined with increased loading from new buildings or removal of support at the toe (resulting from road widening or other construction work). Stability can thus be significantly improved by installing drainage paths to reduce the destabilising forces. Once the slip has occurred, however, a weakness along the slip circle remains, which may then recur at the next monsoon.

Slope stability issues can be seen with almost any walk down a ravine in an urban setting. An example is shown earlier, where a river is eroding the toe of a slope, and there is a swimming pool near the top of the slope. If the toe is eroded too far, or the swimming pool begins to leak, the forces driving a slope failure will exceed those resisting failure, and a landslide will develop, possibly quite suddenly.

Slope Stability Analysis

Slope stability analysis is performed to assess the safe design of a human-made or natural slopes (e.g. embankments, road cuts, open-pit mining, excavations, landfills etc.) and the equilibrium conditions. The term slope stability may be defined as the resistance of inclined surface to failure by sliding or collapsing. The main objectives of slope stability analysis are finding endangered areas, investigation of potential failure mechanisms, determination of the slope sensitivity to different triggering mechanisms, designing of optimal slopes with regard to safety, reliability and economics, designing possible remedial measures, e.g. barriers and stabilization.

Successful design of the slope requires geological information and site characteristics, e.g. properties of soil/rock mass, slope geometry, groundwater conditions, alternation of materials by faulting, joint or discontinuity systems, movements and tension in joints, earthquake activity etc. Choice of correct analysis technique depends on both site conditions and the potential mode of failure, with careful consideration being given to the varying strengths, weaknesses and limitations inherent in each methodology.

Before the computer age stability analysis was performed graphically or using hand-held calculator. Today engineers have a lot of possibilities to use analysis software, ranges from simple *limit equilibrium* techniques through computational limit analysis approaches (e.g. Finite element limit analysis, Discontinuity layout optimization) to complex and sophisticated *numerical solutions* (finite-/distinct-element codes). The engineer must fully understand limitations of each technique. For example, limit equilibrium is most commonly used and simple solution method, but it can become inadequate if the slope fails by complex mechanisms (e.g. internal deformation and brittle fracture, progressive creep, liquefaction of weaker soil layers, etc.). In these cases more sophisticated numerical modelling techniques should be utilised. In addition, the use of the risk assessment concept is increasing today. Risk assessment is concerned with both the consequence of slope failure and the probability of failure (both require an understanding of the failure mechanism).

Within the last decade (2003) Slope Stability Radar has been developed to remotely scan a rock slope to monitor the spatial deformation of the face. Small movements of a rough wall can be detected with sub-millimetre accuracy by using interferometry techniques.

Conventional Methods of Analysis

Most of the slope stability analysis computer programmes are based on the limit equilibrium concept for a two- or three-dimensional model. In rock slope stability analysis conventional methods can be divided into three groups: kinematic analysis, limit equilibrium and rockfall simulators.

Limit Equilibrium Analysis

The conventional limit equilibrium methods investigate the equilibrium of the soil mass tending to slide down under the influence of gravity. Transitional or rotational movement is considered on assumed or known potential slip surface below soil or rock mass. In rock slope engineering, methods may be highly significant to simple block failure along distinct discontinuities. All methods are based on comparison of forces (moments or stresses) resisting instability of the mass and those that causing instability (disturbing forces). Two-dimensional sections are analyzed assuming plain strain conditions. These methods assume that the shear strengths of the materials along the potential failure surface are governed by linear (*Mohr-Coulomb*) or non-linear relationships between shear strength and the normal stress on the failure surface. Analysis provides a factor of safety, defined as a ratio of available shear resistance (capacity) to that required for equilibrium. If the value of factor of safety is less than 1.0, slope is unstable. The most common limit equilibrium techniques are methods of slices where soil mass is discretized into vertical slices. Results (factor of safety) of particular methods can vary because methods differs in assumptions and satisfied equilibrium conditions.

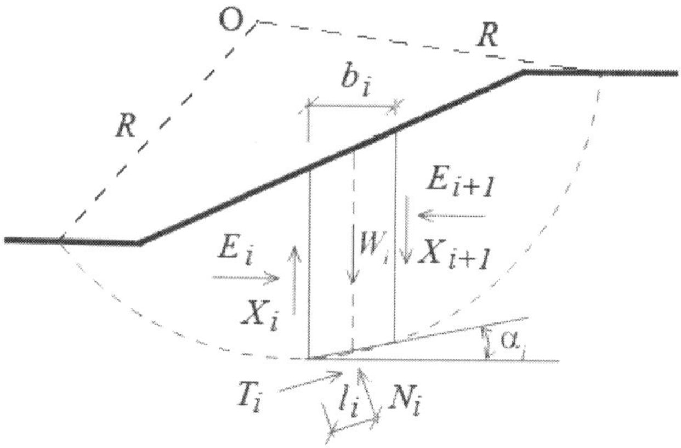

Figure: *Method of slices*

Functional slope design considers calculation with the *critical* slip surface where is the lowest value of factor of safety. Locating failure surface can be made with the help of computer programmes using search optimization techniques. Wide variety of slope stability software using limit equilibrium concept is available including search of critical slip surface. The programme analyses the stability of generally layered soil slopes, mainly embankments, earth cuts and anchored sheeting structures. Fast optimization of circular and polygonal slip surfaces provides the lowest factor of safety. Earthquake effects, external loading, groundwater conditions, stabilization forces (i.e. anchors, georeinforcements etc.) can be also included. The software uses solution according to various methods of slices, such as *Bishop simplified*, *Ordinary method of slices* (*Swedish circle method/Petterson/Fellenius*), *Spencer*, *Sarma* etc.

Sarma and *Spencer* are called as rigorous methods because they satisfy all three conditions of equilibrium: force equilibrium in horizontal and vertical direction and moment equilibrium condition. Rigorous methods can provide more accurate results than non-rigorous methods. *Bishop simplified* or *Fellenius* are non-rigorous methods satisfying only some of the equilibrium conditions and making some simplifying assumptions.

Another limit equilibrium programme SLIDE provides 2D stability calculations in rocks or soils using these rigorous analysis methods: *Spencer*, *Morgenstern-Price/General limit equilibrium*; and non-rigorous methods: *Bishop simplified*, *Corps of Engineers*, *Janbu simplified/corrected*, *Lowe-Karafiath* and *Ordinary/Fellenius*. Searching of the critical slip surface is realized with the help of a *grid* or as a *slope search* in user-defined area. Programme includes also probabilistic analysis using *Monte Carlo* or *Latin Hypercube simulation* techniques where any input parametre can be defined as a random variable. Probabilistic analysis determine the probability of failure and reliability index, which gives better representation of the level of safety. *Back analysis* serves for calculation of a reinforcement load with a given required factor of safety. Programme enables *finite element* groundwater seepage analysis.

Programme SLOPE/W is formulated in terms of moment and force equilibrium factor of safety equations. Limit equilibrium methods include *Morgenstern-Price*, *General limit equilibrium*, *Spencer*, *Bishop*, *Ordinary*, *Janbu* etc. This programme allows integration with other applications. For example *finite element* computed stresses from SIGMA/W or QUAKE/W can be used to calculate a stability factor by computing total shear resistance and mobilized shear stress along the entire slip surface. Then a local stability factor for each slice is obtained. Using a *Monte Carlo* approach, programme computes the probability of failure in addition to the conventional factor of safety.

STABL WV is a limit equilibrium-based, Windows software based on the stabl family of algorithms. It allows analysis using Bishop's, Spencer's and Janbu's method. Regular slopes as well as slopes with various types of inclusions may be analyzed.

SVSlope is formulated in terms of moment and force equilibrium factor of safety equations. Limit equilibrium methods include *Morgenstern-Price*, *General limit equilibrium*, *Spencer*, *Bishop*, *Ordinary*, *Kulhawy* and others This programme allows integration with other applications in the geotechnical software suite. For example *finite element* computed stresses from SVSolid or pore-water pressures from SVFlux can be used to calculate the factor of safety by computing total shear resistance and mobilized shear stress along the entire slip surface. The software also utilizes *Monte Carlo*, *Latin Hypercube*, and the APEM probabilistic approaches. Spatial variability through random fields computations may also be included in the analysis.

Some other programmes based on limit equilibrium concept:

- GALENA - includes stability analysis, back analysis, and probability analysis, using the *Bishop*, *Spencer-Wright* and *Sarma* methods.
- GSLOPE - provides limit equilibrium slope stability analysis of existing natural slopes, unreinforced man-made slopes, or slopes with soil reinforcement, using *Bishop's Modified method* and *Janbu's Simplified method* applied to circular, composite or non-circular surfaces.

- CLARA-W - three-dimensional slope stability programme includes calculation with the help of these methods: *Bishop simplified, Janbu simplified, Spencer* and *Morgenstern-Price.* Problem configurations can involve rotational or non-rotational sliding surfaces, ellipsoids, wedges, compound surfaces, fully specified surfaces and searches.
- TSLOPE3 - two- or three-dimensional analyses of soil and rock slopes using *Spencer method.*

Rock slope stability analysis based on limit equilibrium techniques may consider following modes of failure:

- *Planar failure* -> case of rock mass sliding on a single surface (special case of general *wedge* type of failure); two-dimensional analysis may be used according to the concept of a block resisting on an inclined plane at limit equilibrium
- *Polygonal failure* -> sliding of a nature rock usually takes place on *polygonally-shaped* surfaces; calculation is based on a certain assumptions (e.g. sliding on a polygonal surface which is composed from *N* parts is kinematically possible only in case of development at least *(N - 1)* internal shear surfaces; rock mass is divided into blocks by internal shear surfaces; blocks are considered to be rigid; no tensile strength is permitted etc.)
- *Wedge failure* ->three-dimensional analysis enables modelling of the wedge sliding on two planes in a direction along the line of intersection
- *Toppling failure* -> long thin rock columns formed by the steeply dipping discontinuities may rotate about a pivot point located at the lowest corner of the block; the sum of the moments causing toppling of a block (i.e. horizontal weight component of the block and the sum of the driving forces from adjacent blocks behind the block under consideration) is compared to the sum of the moments resisting toppling (i.e. vertical weight component of the block and the sum of the resisting forces from adjacent blocks in front of the block under consideration); toppling occur if driving moments exceed resisting moments

Stereographic and Kinematic Analysis

Kinematic analysis examines which modes of failure can possibly occur in the rock mass. Analysis requires the detailed evaluation of rock mass structure and the geometry of existing discontinuities contributing to block instability. Stereographic representation (stereonets) of the planes and lines is used. Stereonets are useful for analyzing discontinuous rock blocks. Programme DIPS allows for visualization structural data using stereonets, determination of the kinematic feasibility of rock mass and statistical analysis of the discontinuity properties.

Rockfall Simulators

Rock slope stability analysis may design protective measures near or around structures endangered by the falling blocks. Rockfall simulators determine travel paths and trajectories of unstable blocks separated from a rock slope face. Analytical solution method described by Hungr&Evans assumes rock block as a point with mass and velocity moving on a ballistic trajectory with regard to potential contact with slope surface. Calculation requires two restitution coefficients that depend on fragment shape, slope surface roughness, momentum and deformational properties and on the chance of certain conditions in a given impact.

Programme ROCFALL provides a statistical analysis of trajectory of falling blocks. Method rely on velocity changes as a rock blocks roll, slide or bounce on various materials. Energy, velocity, bounce height and location of rock endpoints are determined and may be analyzed statistically. The programme can assist in determining remedial measures by computing kinetic energy and location of impact on a barrier. This can help determine the capacity, size and location of barriers.

Rock Mass Classification

Various rock mass classification systems exist for the design of slopes and to assess the stability of slopes. The systems are based on empirical relations between rock mass parametres and various slope parametres such as height and slope dip.

Numerical Methods of Analysis

Numerical modelling techniques provide an approximate solution to problems which otherwise cannot be solved by conventional methods, e.g. complex geometry, material anisotropy, non-linear behaviour, in situ stresses. Numerical analysis allows for material deformation and failure, modelling of pore pressures, creep deformation, dynamic loading, assessing effects of parametre variations etc. However, numerical modelling is restricted by some limitations. For example, input parametres are not usually measured and availability of these data is generally poor. Analysis must be executed by well trained user with good modelling practise. User also should be aware of boundary effects, meshing errors, hardware memory and time restrictions. Numerical methods used for slope stability analysis can be divided into three main groups: continuum, discontinuum and hybrid modelling.

Continuum Modelling

Modelling of the continuum is suitable for the analysis of soil slopes, massive intact rock or heavily jointed rock masses. This approach includes the *finite-difference* and *finite element* methods that discretize the whole mass to finite number of elements with the help of generated mesh. In *finite-difference* method (FDM) differential equilibrium equations (i.e. strain-displacement and stress-strain relations) are solved. *finite element* method (FEM) uses the approximations to the connectivity of elements, continuity of displacements and stresses between elements. Most of numerical codes allows modelling of discrete fractures, e.g. bedding planes, faults. Several constitutive models are usually available, e.g. elasticity, elasto-plasticity, strain-softening, elasto-viscoplasticity etc.

Discontinuum Modelling

Discontinuum approach is useful for rock slopes controlled by discontinuity behaviour. Rock mass is considered as an aggregation of distinct, interacting blocks subjected to external loads and assumed to undergo motion with time. This methodology is collectively called the *discrete-element* method (DEM). Discontinuum modelling allows for sliding between the blocks or particles. The DEM is based on solution of dynamic equation of equilibrium for each block repeatedly until the boundary conditions and laws of contact and motion are satisfied. Discontinuum modelling belongs to the most commonly applied numerical approach to rock slope analysis and following variations of the DEM exist:

- distinct-element method
- discontinuous deformation analysis (DDA)
- particle flow codes

The *distinct-element* approach describes mechanical behaviour of both, the discontinuities and the solid material. This methodology is based on a force-displacement law (specifying the interaction between the deformable rock blocks) and a law of motion (determining displacements caused in the blocks by out-of-balance forces). Joints are treated as [boundary conditions. Deformable blocks are discretized into internal constant-strain elements.

Discontinuum programme UDEC (Universal distinct element code) is suitable for high jointed rock slopes subjected to static or dynamic loading. Two-dimensional analysis of translational failure mechanism allows for simulating large displacements, modelling deformation or material yielding. Three-dimensional discontinuum code 3DEC contains modelling of multiple intersecting discontinuities and therefore it is suitable for analysis of wedge instabilities or influence of rock support (e.g. rockbolts, cables).

In *discontinuous deformation analysis* (DDA) displacements are unknowns and equilibrium equations are then solved analogous to *finite element* method. Each unit of *finite element* type mesh represents an isolated block bounded by discontinuities. Advantage of this methodology is possibility to model large deformations, rigid body movements, coupling or failure states between rock blocks.

Discontinuous rock mass can be modelled with the help of *distinct-element* methodology in the form of *particle flow* code, e.g. programme PFC2D/3D. Spherical particles interact through frictional sliding contacts. Simulation of joint bounded blocks may be realized through specified bond strengths. Law of motion is repeatedly applied to each particle and force-displacement law to each contact. *Particle flow* methodology enables modelling of granular flow, fracture of intact rock, transitional block movements, dynamic response to blasting or seismicity, deformation between particles caused by shear or tensile forces. These codes also allow to model subsequent failure processes of rock slope, e.g. simulation of rock

Hybrid/Coupled Modelling

Hybrid codes involve the coupling of various methodologies to maximize their key advantages, e.g. *limit equilibrium* analysis combined with *finite element* groundwater flow and stress analysis adopted in the SVOFFICE or GEO-STUDIO suites of software; coupled *particle flow* and *finite-difference* analyses used in PF3D and FLAC3D. Hybrid techniques allows investigation of piping slope failures and the influence of high groundwater pressures on the failure of weak rock slope. Coupled *finite-/distinct-element* codes, e.g. ELFEN, provide for the modelling of both intact rock behaviour and the development and behaviour of fractures.

Slope Stability Radar

The Slope Stability Radar (SSR) is the application for the monitoring of slope stability at open-cut mines. It is a system currently in use across the global mining and civil industries. Slope stability is a critical safety and production issue for open-cut mines around the world to understand the probability of mine wall failures. Many fatalities have occurred in mining due to slope failure - sudden rock and wall collapses. Even when there is no injury, there is a high cost due to lost production and often damaged equipment. A common technique to determine slope stability is to monitor the small precursory movements, which occur prior to collapse.

Slope Stability Radar (SSR)

Slope Stability Radar, as a contemporary slope stability analysis tool, has been developed by Ground Probe to remotely scan a rock slope to monitor the spatial deformation of the face. Small movements of a rough wall can be detected with sub-millimetre accuracy by using interferometry techniques. The effects of atmospheric variations and spurious signals can be reduced via signal processing means. The advantage of the slope stability radar over other monitoring techniques is that it provides full area coverage without the need for mounted reflectors or equipment on the wall. In addition, the radar waves adequately penetrate through rain, dust and smoke to give reliable measurements, in real-time and twenty-four hours a day.

A series of time measurements over time can be used to track movement of the rock slopes, and also detect variations in the rate of movement, which indicate instability. The radar information can be combined with various other information on a graphical display. Typical examples are 3-D models of the rock surface or pictures of the slope being monitored.

SSR systems are used by mining companies of all sizes for active monitoring of critical slopes and background monitoring of general slopes. It combines the strength of both radar and photographic information to provide a full picture of how a slope is moving. The system uses radar to remotely measure the movement of wall surfaces and uses visual images to confirm and display the result. These measurements enable Geotechnical engineers and mine personnel to track slope movements and set alarms to improve safety and optimize productivity.

SSR systems are operated in the most extreme global conditions and diverse applications of open-pit mining. SSR is used in 19 different countries which include Australia, South Africa, North America, South America, China, Indonesia, India etc.

Remote Sensing Techniques Applied to Landslide Studies

Landslide studies can be organized into three phases

1) detection and classification of landslides
2) monitoring activity of existing landslides
3) analysis and prediction of slope failures in space (spatial distribution) and time (temporal distribution).

Remote sensing techniques can be and are often used in all three stages of a landslide investigation.

Examples of some remote sensing tools (not all of these will be discussed in this page):

- Aerial photography
- Satellite and RADAR images
- Global positioning system (GPS)
- Simulation models and data-bases
- Interferometry
- Stereophotogrammetry
- Soft copy photogrammetry
- Digital photogrammetry

Stage One: Detection and Classification of Landslides

In order to study a landslide, you need to be able to view the size and contrast ofthe landslide features and the morphological expression of the topography within and around the landslide. Parametres you might be interested in determining are the type of movement that has occurred, the degree of present activity of the landslide, and the depth to which movement has occurred.

The most common remote sensing tools used for the detection and classification of landslides are aerial photographs. Aerial photography has become a standard tool aiding in the study of landslides.

Other types of remote sensing techniques used in the detection and classification:

- Stereoscopic imagery
- Satellite imagery
- RADAR imagery

Stage Two: Monitoring Landslide Movement

The second stage in a landslide study is typically monitoring the movement of a landslide. This involves the comparison of landslide conditions over time, including the aerial extent of the landslide, the speed of movement, and the change in the surface topography. There are several remote sensing techniques that are commonly used in this stage of a landslide investigation.

Remote sensing techniques:

- Global Positioning System (GPS)
- Satellite imagery
- RADAR imagery
- Stereophotogrammetry
- Soft copy photogrammetry

Global Positioning System (GPS)

GPS is a useful tool for detecting first stage disaster and further mitigation. It can detect movement of cm/yr, and aid in determining the boundary of the landslide area. Monitors can be placed anywhere you can access, and they are relatively easy to operate. There are some significant disadvantages involved in using GPS, however. Several of these disadvantages are precision is affected by the number of observable satellites present, the obstruction of the observation point, and the monitoring of installed GPS receivers which have been placed out in the field.

Satellite Imagery/RADAR Imagery

There are many types of visible band satellite imagery that can help detect where slides have occurred or where they are about to occur. Change detection images can be taken before and after movement occurred. From these images you can often see surface disruption. Radiometric and geometric calibration is required of the before and after images to make images that have changed stand out from those that have not.

Stereophotogrammetry

In the method of stereophotogrammetry, a satellite acquires two images of the same ground scene within a relatively short period of time, so that it can view surface features have not significantly changed. These images can be processed to get topography from the stereo pair of images. The series of stereo pairs offers a 3 dimensional evolution of the landslide over time.

Landslide Hazard Analysis

Hazard: The probability of occurrence of a potential phenomenon within a specified period of time and within a given area (Varnes, 1984).

Zonation: The division of land into homogeneous areas or domains and their ranking according to degrees of actual/potential hazard caused by mass movements (Varnes, 1984).

Landslide hazard is typically depicted on maps which show spatial distribution of hazard classes, or ýÿlandslide hazard zonations.ýÿ The development of these zonations requires knowledge of the processes active in the area being analyzed, and factors (geologic and triggering) leading to the occurrence of landslides.

Landslide hazard maps typically aim to predict where failures are likely to occur without any clear indication of when they are likely to occur. They are useful for providing landslide hazard information needed for planning and protection purposes. An ideal map of slope instability hazard should reflect information concerning at least the following variables:

- Spatial probability
- Temporal probability
- Type of landslides
- Magnitude of landslides
- Velocity of landslides
- Runout distance of landslides

Temporal probability requires the analysis of potential triggering factors (earthquakes, threshold precipitation, etc.) in relation to the landslides and the application of complex prediction models. Triggering factors, however, cannot be predicted. For this reason, most landslide hazard models typically show spatial probability without regard for temporal probability.

Input data required in physically-based prediction models often varies from location to location, which causes problems when trying to develop a model for a large area. Some alternatives to landslide hazard maps are listed below, but it is in no way a comprehensive list of the numerous methods available for predicting landslide occurrence.

Landslide Inventory Map

A typical landslide inventory map is based on aerial photograph interpretation, ground survey, and/or a database of historical movements within the area. The product of a landslide inventory map is a spatial distribution of landslides as points or to scale. Landslide inventory maps can and often are used as a basis for other landslide hazard zonation techniques or an elementary form of a hazard map. These maps, however, only provide information for a short period of time, and they provide not insight into temporal changes in landslide distribution.

Landslide Activity Map

A landslide activity map is a refinement of the landslide inventory map. It is based almost solely on multitemporal aerial photograph interpretation.

Geomorphological Method Map

A Geomorphological method map is produced by mapping movements in their geomorphological setting. This can be done from aerial photos, satellite/RADAR images, etc. A degree of hazard is allocated at each site in the terrain, but the rules of decision change from place to place (and from person to person), which often leads to a map which is very subjective.

Statistical Techniques

Statistical techniques aim at a higher degree of objectivity and better reproducibility of hazard zonation. There are many statistical methods used in this field.

GIS in Analysis and Prediction of Landslides

A large database is necessary for the analysis and prediction of slope failures. It needs to be able to store, manipulate, and apply the data collected in first two stages (recognition and monitoring). A Geographical Information System is ideal for this stage in a landslide investigation because it is capable of handling large amounts of past, present and future data and integrating this data with predictions. It is capable of data storage and visualization, it is cheaper and easier to use than a manual map production and overlay, it can have regional databases, and therefore, perform both local and regional modeling. There are many types of GIS packages which differ in terms of hardware requirements, potential of spatial functions, efficiency of the data-base, and internal data structure.

Remote sensing techniques greatly aid in the investigations of landslides, on both a local and regional scale. Although they do not replace fieldwork, interdisciplinary research strategies, and testing the reliability of landslide prediction models, remote sensing techniques do offer an additional tool from which we can extract information about landslide causes and occurrences. Most importantly, they greatly aid in the prediction of future landslide occurrences, which is very important to those who reside in areas surrounded by unstable slopes.

Landslide Analysis in Geographic Information Systems

Geographic Information Systems (GIS) have become the promising tool for an effective analysis associated with the study of geologic hazards. GIS is an ideal tool for landslide modeling owing to its versatility in handling a large set of data, providing an efficient environment for analysis and display of results with its powerful set of tools for collecting, storing, retrieving, transforming and displaying spatial data from the real world. This paper demonstrates the ability of the GIS to incorporate the spatially varying data of ground elevation, soil properties, etc. in the engineering analysis of the slope stability. The key factor in landslide hazard mapping is the assessment of risk associated with the failures. Though in general, models are available, better interpretation and understanding of the failures could be derived from them in a GIS environment. A typical landslide area at Mangalore is under investigation for its hazard potential mapping. This paper discusses the analysis of the earthquake and rainfall induced slope failures in GIS and formulates the possible framework.

Landslides constitute one of the major hazards that cause losses in lives and property. Landslides are one of the complex analyses, involving multitude of factors and need to be studied systematically in order to evaluate the hazard. The increasing computer-based tools are found to be useful in the hazard mapping of landslides. One of such significant tools for hazard mapping of landslides is Geographic Information Systems (GIS). A GIS is defined as a powerful set of tools for collecting, storing, retrieving at will, displaying, and transforming spatial data (Burrough and McDonnel, 1998). One of the main advantages of the use of this technology is the possibility of improving hazard occurrence models, by evaluating their results and adjusting the input variables. An important aspect of landslide investigations is the possibilities to store, treat, and analyze spatiotemporal data that are available.

GIS and Landslide Analysis

The unique capability of GIS to work (to capture, store and manage the data) with data referenced by vast spatial or geographic coordinates and its ability to incorporate appropriate engineering models, have caused its proliferating application across the wide sections of engineering, especially in civil engineering where, management of spatial-data is pivotal for the analysis. As the typical landslide analysis demands, collection of numerous data, storage of them and using them in the analysis could be handled well in the GIS environment. Any spatially-distributed data with a geo-reference to real world could be stored as points, lines and polygons (vector model) or as continuous fields (raster data model). Beyond GIS being used as a spatial database, it assists in modelling applications through handling a special form of data that would otherwise be compromised in conventional analysis (Miles et al. 1999). Also, GIS does not only serve as a database for parametre data, rather qualitative and quantitative data can be integrated through spatial relationships rather than through relationships between attributes that may not exist (Frost et al. 1997). The other facilities such as: Query languages and user interfaces permitting rapid modification of parametre values; Convenient and quick updating of model parametres; Overlay function, where multiple maps are either visually or topologically combined and the its potential in visualization of data using the graphic features, that assists the engineer in verifying data and information pertaining to the model and its application; and developing elevation maps and subsequent slope, aspect and hillshade themes (which are useful in the landslide analysis) are worth mentioning, as they are common to many applications of engineering models (Miles et al. 1999). In particular, the ability of GIS to present the data and analysis results in map forms plays a key role in identifying the critical areas (where more rigorous analysis and improved solution is required) by its interactive visualization in a spatially optimized mode.

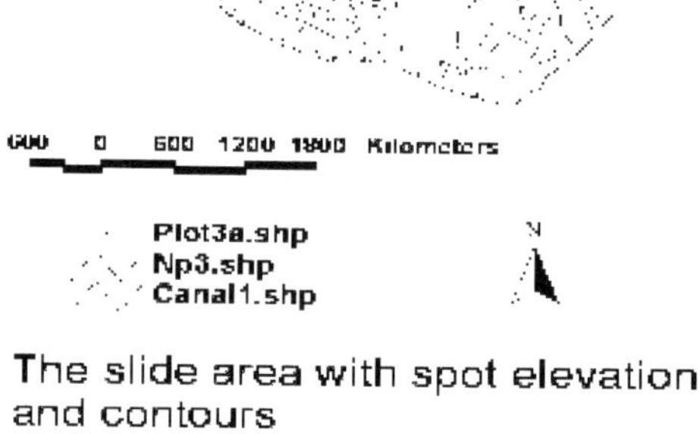

The slide area with spot elevation and contours

Figure: *Landslide area at Mangalore, Karnataka*

Hazard Mapping of Landslides

Landslide hazard maps have been constructed by different methods, such as, from inventories; by consideration of site conditions including geology, hydrology, topography, and/or geomorphology; by statistical correlation of landslide frequency with geologic and geomorphic factors; using safety factors from stability analysis. Mechanics based models have also been used to estimate failure probability based on uncertainties about infiltration of rainfall, or pore pressure, and soil strength (Wu et al. 2000). An a priori study of causal factors of landslides indicate, the ground conditions, geomorphological processes, physical processes and man-made processes as the significant contributors. The pertaining data needed for landslide analysis for a particular slope are:

- Geologic and Geomorphologic features/setting
- Types and quantification of soil and/or rock properties
- Details of vegetation cover, folds, faults, etc.
- Past records of rainfall and earthquake incidences
- Appropriate hydrologic and stability models

Formulation of Typical Landslide Analysis

As part of the research project, a landslide at Mangalore, Karnataka is in investigation. A slope stability analysis combined with risk analyses is proposed to carry out. The area under investigation is shown, marked with 100 point elevations and contours of 2m interval. It could observed that the hill is sloping downward towards southeast. The slide has occurred at the upper regions and hence short-term remediation in the form of relief wells and reinforcing piles were employed at appropriate locations. Various man-made features are also present, like roads and drainage structures, a stream falling in the downslope. However, the theoretical framework for the hazard mapping of the slide area presented has been done in detail. The steps involved in the landslide hazard mapping are given in figure depicted earlier, as flow chart. The detailed formulation of the work is described below. Further, an illustration with a hypothetical slope for simulated soil and earthquake conditions is demonstrated at the end. The methodology proposed would be able to develop the hazard maps, incorporating the triggering causal factors, viz., both rainfall and earthquakes.

Site Conditions

The ground surface elevations could be taken from the topographic maps or digital elevation models (DEM). The other input parametres for the predictive model, involves the detailed characterization of the soil conditions, viz., the description of soil strength properties (cohesion, angle of internal friction, unit weight), permeability characteristics, depth of soil cover, vegetation pattern. The soil cover depth is important in case of shallow soil sliding, which occurs prevalently. The presence of vegetation covers influences the stability calculations significantly as the forested area has increased cohesion value of the soil than the clear-cut areas. Pertinent literature shows that forested areas has upto 20% increased cohesion value (Wu et al. 2000). The values of strength properties would eventually vary spatially throughout the slope. If considerable effort is paid in detailed site investigation, in determining the strength values at more number of points, then a separate statistical analysis (Geostatistics or random field analysis) could be performed, to evaluate the correlation distance. With the information of the correlation distance of each property varying in both vertical and horizontal directions, the interpolation of strength values at unsampled points could be done in Arc View through the Kriging operation. The Arc View facilitates the user to write an avenue coding through which this operation could be performed.

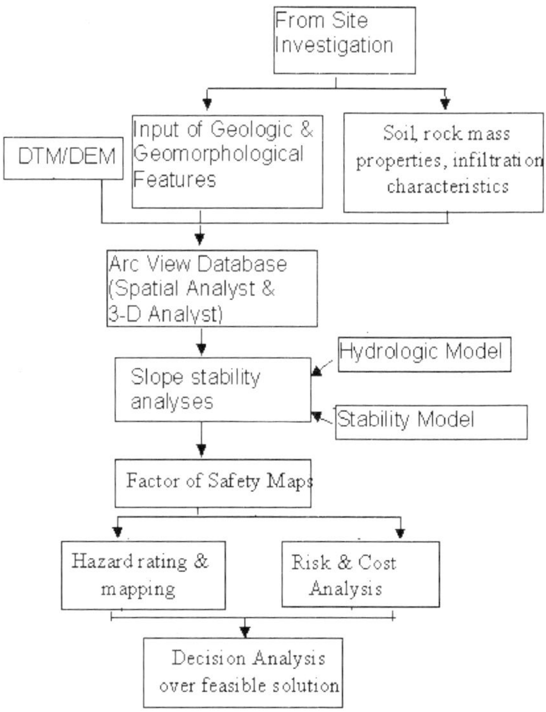

***Figure:** Formulation of landslide hazard mapping*

The geostatistical interpolation is more accurate than the available IDW method or spline methods of interpolation, especially for the geological properties.

Infiltration and Drainage Model

The rainfall-induced failures are the common occurrences in soil slopes. Depending upon the intensity and duration of rainfall and permeability characteristics of the slope, infiltration causes development of pore pressures. Rise in piezometric level in hillside slopes, decreases the effective stress, eventually leading to failure of soil mass. Hence prediction of piezometric levels constitutes an important element in the evaluation of landslide hazards. Suitable hydrologic model predicting the level of pore pressures induced for a particular intensity of rainfall and for the given soil, hydraulic and slope conditions should be coupled with the stability model. The lumped parametre model given by Wu et al (1996), gives the recharge of the groundwater as a function of rainfall characteristics and soil properties. Then a finite-difference solution should be used to calculate the pressure head at different points in a slope.

An Illustrative Example

The preliminary analysis of the slope stability is done in Arc View GIS, with hypothetical slope and simulated earthquake conditions, to validate and explore the capabilities of GIS. The area of analysis is 80X80m, with the crest of 85m and mean slope as 53°. The slope, soil properties (cohesion, friction angle, and unitweight) are given as input in attribute table and developed as themes through event theme menu. Then interpolation for the elevation and soil properties for whole slope area is done through spline interpolation. The following table gives the statistics of the given inputs.

Table: *Input Parametrees for the hazard predition model*

Statistics	*c (kPa)*	*f degree)*	*g(KN/m)*	*Slope (°)*
Mean	38.4	46.2	18.12	53.4
Standard Deviation	1.9	1.467	0.86	16.57
Coefficient ofVariation (%)	5	3	4.7	31
Minimum value	28	41	16	-
Maximum value	42	49	21	-

In stability calculations, it is assumed the failure is planar surface and hence equations 2 and 3 are used for computing safety factors. As this stability model is capable of including the earthquake effects, a pseudo-dynamic analysis is also performed. The factors of safety maps/contours with hazard rating are shown in figures. Assuming that the probability distribution of factors of safety follows normal distribution, the reliability index and the corresponding probabilities of failure are computed, for both static and pseudo-static results.

Table: *Results of static and pseudo-static analyses*

	Static FS	***Dynamic Factors of Safety (Aah)***				
		0.1	0.2	0.3	0.4	0.5
Mean of FS	2.6	1.95	1.52	1.21	0.98	0.8
s of FS	0.23	0.19	0.19	0.19	0.19	0.21
C.O.V (%)	8.8	37	12.5	15.8	20.3	26
b	6.89	5	2.73	1.11	-0.01	-0.97
P[f]	0	0	0.02	0.11	1.5	1.54

The safety factor maps gives the visualization of the weaker and stable zones, spatially. Accordingly, the distribution of hazard areas could be spotted for further in-depth analysis or for suitable remediation programme. Relevant maps and contours are given in Figures.

$$b = \frac{E[F] - 1.0}{\sigma_F} \quad (5)$$

$$P[F] = 1.0 - f(b) \quad (6)$$

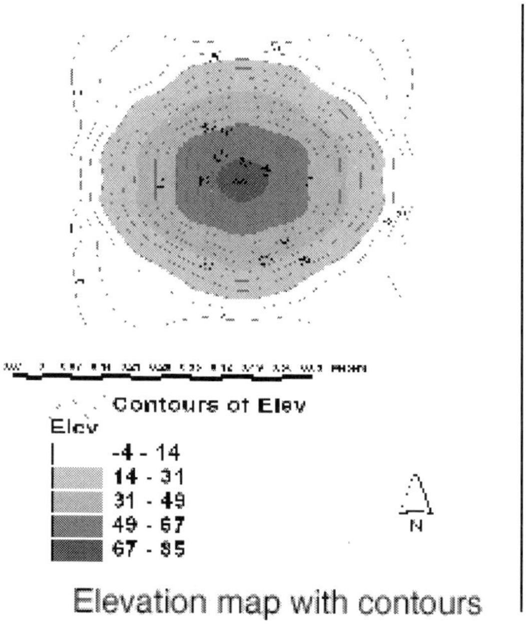

Figure: *Elevation map with contours of the slope*

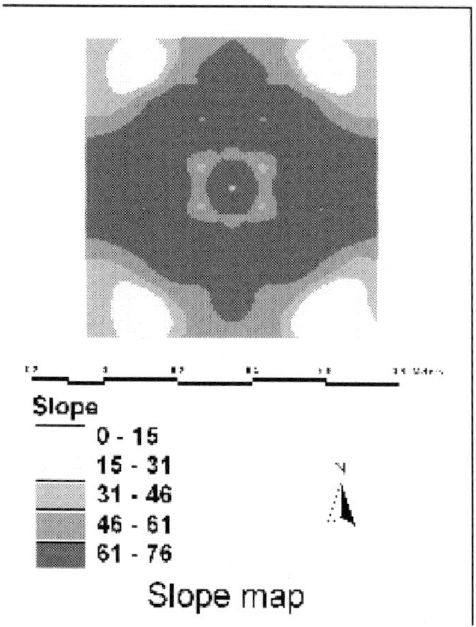

Figure: *Slope map of the analysis area*

Concluding Remarks

Geographic Information Systems (GIS) is being exploited widely in many engineering problems, which involves spatial data management. Landslide hazard mapping, one of the important task in disaster/hazard mitigation projects is a typical problem involving huge database. A framework for the landslide hazard mapping in GIS is discussed with an illustrative example. The posterior analysis of GIS results gives the engineer a better understanding and visualization of the problem and results.

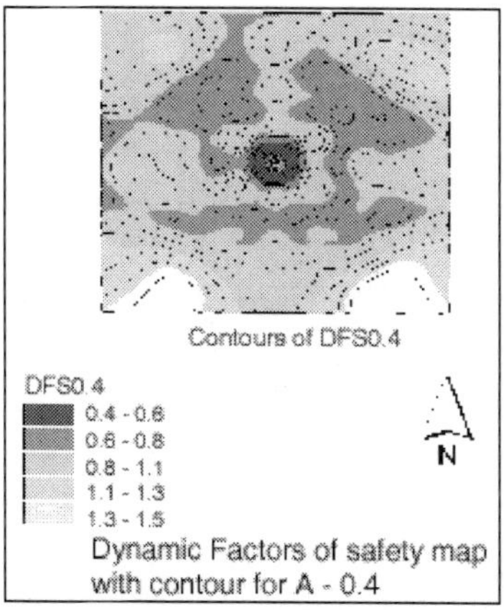

Figure: *Static factors of safety of the slope*

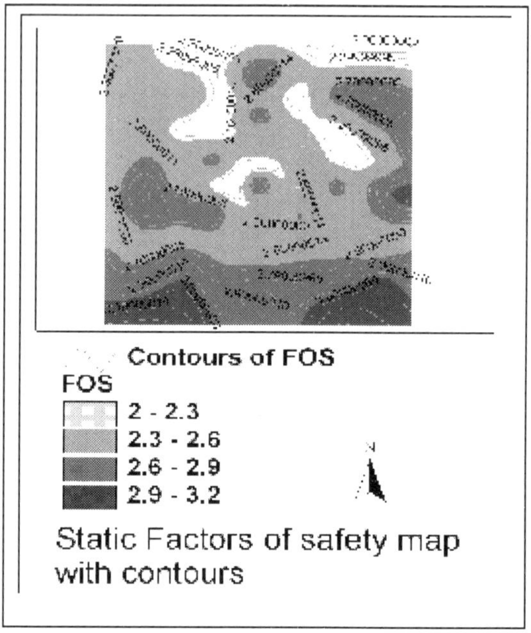

Figure: *Dynamic factors of safety for the slop with acceleration coefficient of earthquake, 0.4.*

Landslide Investigation and Prediction

The flow chart shown in figures, describes the general investigation procedures in an attempt to understand the mechanism of origination of disasters associated with slope movement and to predict the resulting deformation. Investigation items and investigation methods are shown on Table above.

Preliminary Investigation

Collection of Existing Data, Data Review: Landslides often occur at specific locations under certain topographic and geologic conditions. Therefore it is important to utilize existing data (history of the problem, records of restoration work, and data review) in order to understand the topography, geology, and properties of similar landslides.

It is also important to understand their relationship with meteorologic factors, period of activity, existence of any warning sign, ground water conditions, chronology of topographic change or erosion by rivers, earthquakes, and other factors which may have a relationship with the slope deformation surrounding the investigation site area prior to the detailed investigation.

Topographic Investigation: It is necessary to identify any changes in the site topography. That can be accomplished by recognizing; 1) the overall topographic feature of the site; 2) understanding the topographic characteristics of the site slopes; and 3) estimating the regional geologic structure of the site. Such methods include comparing the aerial photographs of the site and vicinity taken prior to and after the sliding, and interpreting the topographic maps and aerial photographs.

In Japan, aerial photographs are taken every few years over the entire country at a scale between 1/10000 to 1/40000. These photographs are used to understand the chronologic and topographic changes over the country. Furthermore, in order to be able to effectively interpret the phenomena related to microtopography and landslides, large scale aerial photographs with a scale of 1/8000 to 1/15000 are often taken. By utilizing aerial photographs, it is possible to interpret landslide phenomena and warning signs, geology and geologic structure, topography and distribution of vegetation type. For landslide investigations, it is useful to identify and interpret the distribution and continuity of knick lines, gentle slopes, gullies and cracks in the photos to aid in preparing a photo interpretation map. The map can then be utilized during the field investigation.

The recent popularity of remote sensing using satellite photographs has been particularly useful for analysis utilizing the thermal infrared spectrum which is possible to estimate the distribution of slide areas and ground water, and live vegetation. Remote sensing can be used for analysis of topographic characteristics and topographic in terrain susceptible to landsliding.

Field Investigation: With an approximate understanding of the overall topographic feature and knowledge of the distinction(s) of movement and aerial extent of the sliding block(s) (viewed from the opposite side), a detailed field investigation plan can be developed to delineate the aerial extent and a general direction of movement of the landslide zone, assess the geology and geologic structure, estimate the cause(s) of the sliding, and predict future movement. The field investigation should not include just the actual landslide area, but also exists. The field investigation should also include areas where aerial photographic interpretation is difficult or unclear, and in areas that could aid in the understanding of particular topographic features and characteristics.

Drafting a Detailed Investigation Plan

In order to examine the follow item, a detailed investigation which will satisfy the objectives under the listings in the investigation methods and observation instruments in the Table above, should be planned.

- aerial extent of the slide, differentiation of moving blocks and identification of the direction of movement
- location and shape of slide plane(s)
- nature of landslide block(s)
- possibility of further or future movement on slopes above the existing slide
- possibility of further, future or accelerated sliding
- distribution of ground water

Survey lines can be established on each moving block on the ground where the slide mass is expected to be thickest and where the stability analysis and plan for control works will be emphasized. As a general rule, the main survey line should be placed where the width of the slide exceeds 100m with subsidiary survey lines established at approximately 50m intervals.

Exploratory borings should be drilled on the order of every 30-40m. At least three borings should be drilled along the main survey line with one boring drilled at least 5 to 10 m below the slide plane. During the early stage of the investigation, it is particularly important to have an accurate estimate of the configuration and location of the slide plane(s) an adequate boring depth can be achieved.

Seismic survey lines should be placed at intervals between 50-100m, and electric survey and radioactivity survey lines should be placed at 20-50m intervals. A survey should be conducted along the main survey line as well as along the longitudinal survey lines that cross the main survey line and subsidiary survey lines. For the seismic survey, the survey points should be established at 5-10m intervals, 20-50m interval for the electric survey, and at 3-5m intervals for the radioactivity survey. Furthermore, to verify the results of the geophysical surveys it is important to drill borings at the survey line intersections.

Detailed Investigation

Investigation of Surface Deformation: The investigation of surface deformation is conducted to define the boundaries of the landslide, size, level of activity and direction(s) of the movement, and to determine individual moving blocks of the main slide. The presence of scarps and transverse cracks are useful for determining whether the potential for future activity exists.

Instrumentation used for the surface deformation investigation includes extensometres, ground tiltmetres, movement determination by survey methods including transverse survey, grid survey, laser survey from the opposite bank, movement determination by aerial photographs, and G.P.S. provides an example of instrumentation.

- Extensometre The extensometre is used to measure relative movement by comparing the extension of two points. The extensometres are generally installed across the main scarp, at transverse crack and transverse ridges near the toe or front portion of the slide and parallel to the suspected slide movement. By arranging a series of interconnecting extensometres from the main scarp to the toe of a complex landslide that has many moving slide blocks, the resulting data could aid in clearly delineating the individual slide blocks. Measurements should be accurate to within 0.2mm, and the magnitude of the movement and daily rainfall data should be included to establish the relationship between the measurable movement and the precipitation rate.
- Tiltmetre The ground tiltmetre is useful for determining the deformation at the head and toe portions and sometimes along the flanks of the landslide, or to assess the possibility of future deformation. A level type tiltmetre is most conventional. The tiltmetre is capable of measuring the N-S and E-W components. The magnitude of tilting and tilt directions can be determined directly from the instrument panel. Furthermore, in order to determine the characteristics of the deformation, the results are shown chronologically along with the daily rainfall totals. The relationship between the magnitude of tilting and the cumulative effect of tilting, rainfall totals and groundwater levels are shown.
- Simple Method to Measure Movements One of the simplest methods to determine landslide movement is to drive wooden stakes across a tension crack along the direction of slide movement. Then attach horizontal boards to the stakes, and saw through the boards. Any movement across the tension crack can be determined by measuring the space between the sawed portions of the boards.

- Determination of Movement by Surveying (Transverse Survey, Grid Survey, Laser Survey From the Opposite Bank, Movement Determination by Aerial Photographs, and G.P.S.).

Transverse Survey: This survey method establishes transverse survey lines across the landslide blocks with closely-spaced survey stakes. The survey stations should be established both within the slide and outside the slide on stable ground.

Grid Survey: This survey method involves constructing grid lines across the entire landslide as well as stable ground outside of the landslide. The survey stakes are driven at the intersection of the grid lines.

Laser Survey from the Opposite Bank: A control point is established along the opposite bank on stable ground, and survey stakes are positioned within the slide. It is most effective where the movement in large.

Movement Determination by Aerial Photographs: For landslides with a large component of movement, aerial photographic determination is the most useful. An accurate movement can be measured by annual or bi-annual flying.

G.P.S.: Global Positioning System is the state of the art technology that uses signals from satellites to determine the three-dimensional positioning of the slide. G.P.S. has been used in recent landslide investigations where a high degree of success has been reported.

- Automating Survey System In the past, measurements of slope deformation have been performed manually. Recently, automatic survey systems using data loggers and computers have adopted. The instrument set-up in the field has been designed for easy installation, and is weatherproof, durable, maintenance-friendly and economical. Through remote control in real time and rapid geographical data processing, it is possible to store long term data accurately and effectively and would provide an early warning of slide activity, thereby reducing landslide hazard. Furthermore, the recent development in the informationalized constructions systems and adopting the safety control at the construction site using the real time facilitates the planning through construction stage.

There are Three Main Advantages in Using the Automating Survey System

- Surveillance of the conditions of landslide failure: Issuance and cancellation of landslide watch and warning announcement based on the velocity of movement, piezometric pressure and variation in the rainfall amounts. Prediction and forecasting of the landslide failure.

- Understanding of the conditions of the landslide deformation: Chronological measurements of movement velocity. Determination of the slide plane depth. Determination of the relationship between the slope deformation and factors of slide occurrence (pore water pressure against the slide plane, critical pore water pressure related to the time of sliding, rainfall and snowmelt).
- Effectiveness in determining landslide control works: Measurement of the amount of earth movement and pore water pressure. Measurement of the earth pressure affected by piles and collection wells. Determination of the effectiveness of construction.

Investigation of Geologic Structure

In most cases, the investigation of geologic structure relies on exploratory borings; however in cases where the bedrock distribution is ambiguous or a better understanding of the regional geologic structure is needed, then a geophysical exploration (seismic survey, electrical survey and radioactivity survey) is combined with the boring data.

Borings

The majority of the borings drilled are larger than 66mm. Core samples are recovered from the borings and are stored in core boxes. Boring logs should be prepared along with photographs of the core samples. The boring logs shall include such information as: geologic and soil description; colour; hardness; lithologic description; degree of weathering; alterations and fractures; strike and dip of bedding joints; boring conditions; initial and stabilized ground water levels; and rate of core recovery.

Geologic assessment based on the boring data obtained from the drilling site should include a discussion regarding the differentiation of moving earth blocks, semi-moving earth, and stable ground. Clays within the slide plane generally have a high moisture content, are highly sticky and plastic and are often associated with abrasion scars and slickensides. During drilling, squeezed earth could occur near slide plane. Slopes where advanced relaxation of the bedrock formation has occurred will often exhibit gentler slopes than that of the unaffected bedrock zone. Formations can bend or form a kink bend near the lower limit of this zone, and could develop into a slide plane. In translational dip slope slides, the slide plane in many instances will develop along a thin, weak bed of mudstone, tuff bed or coal seam sandwiched between hard and competent beds. Borings can sometimes easily miss these thin beds. Therefore, the possible existence of slide planes along there weak beds typically consist of about 10cm, and must be considered even though the boring may not indicate they are actually present.

Furthermore, using the data from the borings, the following information must be assessed or determined.

Evaluation of slide plane 2. Ground water level measurements 3.Ground water logging 4. Ground water tracer tests 5. standard penetration tests, Horizontal loading tests, In-situ tests such as in site permeability tests 6. Sampling for soil tests 7. Various geophysical logging.

Geophysical Surveys

Geophysical surveys (seismic survey, electric survey and radioactivity survey) are conducted to understand the approximate geophysical conditions of the slide itself and the surrounding area. P-wave refraction surveys are the most common seismic survey. Other methods, such as S-wave and P-wave shallow refraction, are seldom used. Electric survey is the specific resistance method and is applied to determine the distribution of aquifer(s) and to understand the geologic structure. These surveys include the development of the geotomography method. A natural radioactivity survey is used to determine the locations of small scale fracture zones and cracks.

Evaluation of Slide Plane

Determining the slide plane for actively moving landslides utilize the fact that the rates of movement differ significantly along the slide plane. Depending on the requirements for surveying accuracy and magnitude of movement, the appropriate instrumentation shall be selected from the following representative instruments;

Pipe Strain Gauge 2. Inclinometre 3. Multi-Layer Movement Metre

Pipe Strain Gauge: P.V.C. pipes with strain gauges are inserted into the boreholes, and the movement is estimated by the change in the strain as the P.V.C. pipe bends. The accuracy of the strain gauge increases as the intervals of the gauge narrows, however, it is acceptable to widen the space as much as 1m for investigations involving very thick slide materials and when it is difficult to handle the survey extension wires. Two of the lowest strain gauges must be anchored into the bedrock below the slide plane so that data from within the intact formation can be obtained. Furthermore, annular space between the borehole and the pipe must be filled with concrete following the gauge installation. The instruments should last for one to two years.

Inclinometre

A grooved casing is inserted into the borehole extending into the bedrock formation, and have an adequate quality of grout placed into the borehole to assure a positive contact with the borehole. By lowering a probe equipped with a tilt sensor, deformation in the casing can be detected and movement of a landslide can be determined. An accurate measurement is possible where the deformation of a landslide is relatively small. As a landslide movement increases, the borehole and casing will bend making insertion of the probe difficult or will exceed the tilt detection limit of the instrument.

Multi-Layer Movement Metre

several wires are anchored at different depths within a borehole with the attached wires extended to the ground surface. The magnitude of the displacement of each wire segment can be measured directly using a ruler. It is possible to install 20 to 30 wires per borehole. This method is not suitable for landslides with small displacement. This instrument is most effective where the slide movement is so large that some of the other instruments cannot be used. Applying the same principle, a vertical extensometre can be constructed by fixing a wire on the bedrock at the bottom of the borehole.

Other Methods

Other methods to evaluate the slide plane include: slide plane detection probe; creep wells; and sounding penetration test.

Ground Water Investigation

Investigation of ground water, which is a driving force of sliding, includes determining ground water level, pore water pressure, ground water logging, ground water tracing test, pumping test, water quality analysis, electricity survey, geothermal survey, and geophysical logging (electric logging and radioactive logging). Based on the results of the above measurements and tests, ground water control works can be planned and designed.

Ground Water Level Observation: As a general rule, ground water levels should be measured in all the boreholes. In some of the more important boreholes, continuous rainfall data will be kept by an automatic recorder to determine the correlation between the slide movement and rainfall and ground water level, and will collect data on the ground water distribution and movement regime.

Pore Water Pressure: Ground water levels in boreholes will often reflect seepage from highly fractured formations or indicate the water level of a predominant aquifer. Therefore, for stability analysis, it is best to measure pore water pressure along the slide plane. Sometimes it is difficult to accurately estimate the depth of the slide plane. In such cases it is desirable to install piezometres in the beds with low seepage or low shear strength. The standard piezometres that are used in landslide investigations must be durable, and open piezometre water level type.

Ground Water Logging: Locations of ground water flow and flow directions can be determined by measuring the increasing specific resistance of ground water in flow over time. The measurements will be continued often lowering specific resistance of ground water by injecting a salt solution into the borehole. There should be at least two borings for ground water logging at the head portion of the landslide where abundant ground water is expected. The measurement results should be recorded along with the boring logs, and the relationship between the location of ground water flow and bed, and magnitude and variation of specific resistance of ground water should be discussed. Furthermore, the results of the analysis should be recorded along with the cross sections in order to understand the overall ground water flow.

Ground Water Tracer Tests: Tracers such as a soluble dye, or inorganic chemicals (NaCl) are injected into a borehole. Water samples are then collected chronologically from springs, other boreholes, wells and ponds within or outside the landslide, and are analyzed for the tracer to estimate the ground water flow direction(s) and permeability. This data is used for basic information for the design of dewatering works.

Drawdown Test: In order to estimate the yield and to calculate the coefficient of permeability, water within a borehole is pumped to certain levels after raising the boring casing every 2 to 3m. A time-recovery curve can then be plotted using Jacob's and other formulas, and the coefficient of permeability can be determined.

Water Quality Tests: Water quality tests are an effective method to examine the distribution of the ground water regime and flow directions where the subject landslide is very large and the ground water system is expected to be complicated. Specific tests include determination of water temperature, Cl-, SO2-4, HCO-3, Na+, K+, Ca++, and Mg++ content, pH, alkalinity, electric conductivity, SiO2, and others. The test results are classified according to the analytical data and composition.

Geothermal Investigation: This procedure utilizes ground temperature measurements throughout the study area, including ground temperatures near the ground water veins. By measuring the temperature differences at non-ground water areas and near ground water veins, it is possible to isolate the ground water veins where the temperature difference between the two is large. By conducting the geothermal investigation in summer months or winter months where near surface ground temperature is influenced by air temperature, good results have been obtained for the isolation of relatively shallow ground water.

Geotechnical Investigation (Rock Mechanic Tests)

In order to conduct slope stability analyses and to design appropriate control measures for landslides, physical properties such as strength of slide plane, location and depth of slide plane and stable ground areas must be determined. The following tests are generally performed; physical tests, Standard Penetration Tests, soil mechanic tests (unconfined compression, tri-axial compression, box shear, ring shear, and in-situ shear (along the slide plane)). In order to obtain the earth reaction coefficient for the design of the restraint works, there is a current tendency to conduct more horizontal loading tests and plate loading tests to determine the modules of deformation. Furthermore, the intensity and degree of alteration of the slide plane clays are evaluated by X-ray diffraction methods. The results have also been applied to analyze the origin of the slide plane.

Prediction of Landslide

Landslide Distribution Map: Most of the new landslide are reactivated old failures in landslide terrain, and unless there are special causes, it is extremely rare that non-landslide terrain fails. Those topographic characteristics can be interpreted from aerial photographs and topographic maps, and be verified through field reconnaissance.

Furthermore, bedrock landslides and weathered bedrock landslides with past movement at the time of sliding is small, and sheared bedrock and topographic features related to the early stages of sliding that were subjected to creep deformation in the deeper portions often do not exhibit clear landslide topographic characteristics. Because of these reasons, double ridge topography associated with mountain deformation, parting ridges, breaks-in-slope, knick lines, distribution of old and scarps, bulging at the tip of ridge lines, discrepancy in the geologic distribution "following the investigation, geologic structure degree of shearing degree of creep and other factors must be considered when evaluating landslide topography.

Landslide distribution maps with the above descriptions are generally limited to small areas, however, recent regional maps covering the entire country of Japan have been published.

Landslide Prediction: Now it is possible to predict the timing of a slope failure by interpreting the rate of deflection measured by extensometres placed across tension cracks of a slope. Failure predictions rely on extensometres placed across scarps, and areas will be considered "off-limits" when the rate of movement exceeds 2 to 4 mm/hour. Based on the change in the rate of movement, the following three methods are commonly used to predict the timing of landslide movement.

- Graphic solution using the tertiary creep curve
- Graphic solution of surplus time using semi-log paper
- Method using a inverse number of rate of movement

Landslide Prediction Mapping

Landslide hazard analysis and mapping can provide useful information for catastrophic loss reduction, and assist in the development of guidelines for sustainable land use planning. The analysis is used to identify the factors that are related to landslides, estimate the relative contribution of factors causing slope failures, establish a relation between the factors and landslides, and to predict the landslide hazard in the future based on such a relationship. The factors that have been used for landslide hazard analysis can usually be grouped intogeomorphology, geology, land use/land cover, and hydrogeology. Since many factors are considered for landslide hazard mapping, GIS is an appropriate tool because it has functions of collection, storage, manipulation, display, and analysis of large amounts of spatially referenced data which can be handled fast and effectively. Remote sensing techniques are also highly employed for landslide hazard assessment and analysis. Before and after aerial photographs and satellite imagery are used to gather landslide characteristics, like distribution and classification, and factors like slope, lithology, and land use/land cover to be used to help predict future events. Before and after imagery also helps to reveal how the landscape changed after an event, what may have triggered the landslide, and shows the process of regeneration and recovery.

Using satellite imagery in combination with GIS and on-the-ground studies, it is possible to generate maps of likely occurrences of future landslides. Such maps should show the locations of previous events as well as clearly indicate the probable locations of future events. In general, to predict landslides, one must assume that their occurrence is determined by certain geologic factors, and that future landslides will occur under the same conditions as past events. Therefore, it is necessary to establish a relationship between the geomorphologic conditions in which the past events took place and the expected future conditions.

Natural disasters are a dramatic example of people living in conflict with the environment. Early predictions and warnings are essential for the reduction of property damage and loss of life. Because landslides occur frequently and can represent some of the most destructive forces on earth, it is imperative to have a good understanding as to what causes them and how people can either help prevent them from occurring or simply avoid them when they do occur. Sustainable land management and development is an essential key to reducing the negative impacts felt by landslides.

GIS offers a superior method for landslide analysis because it allows one to capture, store, manipulate, analyze, and display large amounts of data quickly and effectively. Because so many variables are involved, it is important to be able to overlay the many layers of data to develop a full and accurate portrayal of what is taking place on the Earth's surface. Researchers need to know which variables are the most important factors that trigger landslides in any given location. Using GIS, extremely detailed maps can be generated to show past events and likely future events which have the potential to save lives, property, and money.

An Accurate Way of Predicting Landslides

A landslide can seriously injure or even kill people. Now, a new early warning system will be the first to employ geological data in tandem with the latest weather forecasts to provide a concrete warning in emergency situations. A deluge of rain pelts down on the already thoroughly sodden ground. Rivers burst their banks as local residents haul up sandbags to protect themselves from the rising waters. In hilly areas, people are no less anxious, knowing that the waterlogged earth on a hillside can easily slip, burying cars and houses beneath it and, if the worst comes to the worst, even people. Experts use maps marked with danger areas to determine the probability of a specific slope succumbing to a landslide. But these maps only cover a specific point in time, and do not take current weather conditions into account. Of these atmospheric factors, heavy rain in particular can trigger catastrophes.

Now, safety experts are to be supported by the ELDEWAS early warning system. This system makes use of regularly updated weather conditions and forecasts, coupling these to regional information on elevation profiles, slopes and land use, allowing it to issue an early warning in case of danger. ELDEWAS stands for "Early Landslide Detection and Warning System" and is being developed by research scientists at the Fraunhofer Institute of Optronics, System Technologies and Image Exploitation IOSB in Karlsruhe. "The ELDEWAS early warning system goes hand in hand with INCA-CE, a project co-financed by the EU in which researchers are working on improving short-term weather forecasting, or 'nowcasting'," says Dr. Oliver Krol of Fraunhofer IOSB. While standard meteorological data is for the most part only updated once an hour, with a spatial resolution of ten kilometres, the experts working on the INCA-CE project are able to provide weather forecasts at 15-minute intervals with a spatial resolution of one kilometre.

Factoring in Current Weather Conditions

The early landslide warning system is initially being developed by researchers for use in the state of Burgenland in Austria, with the regional safety centre providing all the necessary data. Which slope is this and how steep is it? What type of ground are we dealing with – sand, clay or rock? How is the land used? Where are the installations, houses or roads? Where is wooded and where is open land? Researchers then combine these parametres, which stay stable over the long term, with weather data, which is constantly in flux. The weather data is provided online by Austrian meteorological service ZAMG, which is also taking part in the INCA-CE project. Initial practical testing is planned for the spring, when current weather information will be incorporated into the early warning system for the first time. The prototype is due to be ready in the autumn. "The software will of course also then be available for use in other countries and regions," says Krol. He also explains the aim behind the research, describing how the system will constantly assess the situation in the background until it perceives a danger, at which point it will independently issue a warning comprising the relevant coordinates and the contact details of the person with responsibility for the area concerned. This contact person is then to be automatically warned of the impending incident via text message, giving them time to take appropriate action, evacuating the populace or locking down the area concerned.

Researchers still have a few challenges to overcome before this vision can be realized, however, including how to integrate the online weather data into the system and how to analyze the data received. "There's no doubt that the bulk of the work lies in answering the question of when a situation can be said to become critical. Given that setting rigid threshold values only allows for a binary yes or no answer, offering protection only against the worst case scenario, we have opted to model the problem using fuzzy logic," explains Krol. This means qualifying the threshold values of the various contributory factors, allowing us to link the variables. In this way, we can get as realistic as possible an assessment of the risk posed."

Predicting Landslides

Predicting earthquakes and avalanches is notoriously difficult, scientists have been attempting to do so for hundreds of years with very little success, and a group from imperial college london may have worked out why.

Both earthquakes and avalanches are types of critical phenomina, the classic example is slowly pouring sand onto a sand pile. The top of the sandpile slowly gets more and more unstable until something gives and you get a landslide. The problem is predicting how big the landslide is going to be. It might be tiny or the small landslide, may trigger a larger one which may trigger an even larger one, or it might not. Henrik Jensen has been looking at a simplified version of this. He has been creating a pile of ball bearings by adding one at a time to the top of the pile. Every time he added a ball he took a photo. Occasionally there were landslides of different sizes which he tried to predict in various different ways.

He found that it was virtually impossible to predict the size of the landslide the traditional way, by looking at the size of previous landslide. But he did have more luck when he looked at the state of the pile before each ball was added.

He found that the more disordered the pile was before the next ball was added, the larger the landslip was going to be. and he could predict the size of the landslip with a 64% accuracy, and he thinks that he can get a lot better. This sounds like a pretty abstract finding, but it does show that if you are wanting to predict avalanches or earthquakes you shouldn't be looking at previous avalanches or earthquakes, but look at the state of the hillside or tectonic plates. And possibly more importantly it means that the problem is soluble if we approach it from the right direction.

Listening with Seismology Could Predict Landslides

Landslides carrying the weight of more than 200 million cars careen down mountains about once a month somewhere on Earth. Now, with a global network of seismic monitors, scientists can detect the biggest landslides remotely, according to a study published today (March 21) in the journal Science. The seismic ripples also reveal new information about how landslides work, researchers said.

"The landslides turned out to be not only an interesting curiosity, but we are also discovering things about landslides which were not really observed before," said GöranEkström, lead study author and a seismologist at the Lamont-Doherty Earth Observatory of Columbia University in New York.

The findings are part of an ongoing project to find "strange things" in the unusual wiggles that jog the network's seismometres, Ekström told Our Amazing Planet. Ekström and co-author Colin Stark, also of Lamont-Doherty, sift through long-period seismic waves, which look like relaxed undulations compared to the frantic, high-frequency shaking that results from earthquakes and explosions.

Ekström and Stark spent several years fine-turning a computer algorithm that would automatically extract landslides from the chatter that constantly rattles the Earth. When a promising signal appeared, the researchers would check satellite images, news reports and blogs to verify the shaking was truly a landslide.

Carrying the Critical Data

For a landslide's vibrations to reach remote seismometres by traveling along the Earth's surface, the energy must be as powerful as a magnitude-5.0 earthquake, Ekström said.

The researchers discovered the seismic data carried a landslide's rap sheet — its source, location and weight, along with how far and fast it moved. Surprisingly, that critical data was directly proportional to the length of the mountainside that fell off, or failed, in a landslide, the study found. The relationship was maintained whether a landslide came from a volcano blasting apart, a collapsing rocky cliff or a slurry of typhoon-soaked debris.

> *"This challenges the view that there are many different types of landslides and [that] there are certain characteristics that determine the behaviour of landslides, such as soil type or the type of rock," Ekström said. "There is vibrant research going into understanding the dynamics of how the mass makes it down the mountain and follows the valleys. We're hoping that the fact we show the seismic signal has a lot of information will stimulate this work."*

Landslides are Like Roller Coasters

During a landslide, the cascading debris vibrates the Earth's surface, sending seismic waves through the crust, Ekström explained. The forces follow patterns similar to a roller coaster ride; the rock mass pushes down as it accelerates and drags while it slows down. The shaking is distinct from earthquakes and explosions, which make seismometres jiggle frantically with fast shaking, an initial signal that's over in just a few seconds. Landslides produce long-period undulations on seismograms that last 35 to 150 seconds.

With several seismometres, the researchers can pinpoint where the landslide started, the location of the sliding mass at any point in time and its weight. They determined that a massive slide at Canada's Mt. Steele in 2007 was 100 million tons (10^{11} kilograms), about the weight of all the cars in the United States, according to Ekström. The researchers have also uncovered evidence that some remote landslides that look like one event on satellite images actually consist of several rock falls that collapsed over a period of several days, not all at once.

Predicting Landslides

In the future, seismic detection and analysis of landslides could improve scientist's understanding of how the rock falls work and perhaps provide warning to people living nearby, Ekström said.

"There are regions that have a lot of landslides, like the Himalayas and Hindu Kush, where it would be beneficial to know when things have happened up in the mountains," he said. [Can You Outrun a Natural Disaster?]

One example is a huge landslide above the Seti River in Nepal, which triggered a flash flood on May 5, 2012, causing more than 70 deaths. Ekström and Stark estimated some 780 million cubic feet of rock broke off a ridge near the peak called Annapurna IV, then plunged into the Seti's headwaters at about 160 mph (258 km/h) after traveling about a mile. A pilot in the area witnessed the flood and warned residents, but landslide detection could also provide advance warning.

Ekström said Taiwan is also considering using its seismic network for landslide monitoring in the wake of deadly landslides caused by Typhoon Morakot in August 2009.

"The vibrations in the Earth actually record what these forces are. We are not blind to them. Hopefully, that will help us understand landslides and use that knowledge to predict where and how they will happen," Ekström said.

Landslides Hazards

Earthquakes are a major cause of landslides. Landslides occur when masses of rock, earth material, or debris flows move down a slope due to gravity. Landslides can occur on any terrain if the conditions are right, and cause significant damage and casualties to people and property. In this lesson, students learn about earthquake-induced landslides and the associated hazards, and how and why landslides occur. In addition, students discuss steps they can take to reduce landslide hazards.

Unlike previous lessons, Lesson 8 begins with a Tabletop Exercise. The objective of a Tabletop Exercise is to simulate a complex situation with multiple possible responses before the students have obtained all of the knowledge of the lesson concepts. The historical basis that served as the impetus for the Tabletop Exercise was the 2004 Southeast Asian tsunami, and the story of ten-year-old Tilly Smith. A British tourist traveling with her family in Thailand, she recognized the receding shoreline and strange surface bubbles from a tsunami lesson conducted just weeks before the event. She immediately warned her parents, who warned beachgoers and hotel staff of the impending tsunami. She is directly credited with saving hundreds of lives as a result of her timely and knowledgeable actions.

Ask your students to describe and review what an earthquake is and what causes them. In addition to the direct effects of earthquakes such as ground shaking, ask them to list other hazards associated with earthquakes. Other hazards may include landslides, liquefaction, structural hazards (buildings collapsing), non-structural hazards (a falling bookshelf or a shattered window), and destruction of utility lines (gas, water, electricity) and roads.

Tell your students they are going to learn about earthquake-induced landslides and related hazards in this lesson. Show students pictures of landslides and ask them to list what the photos share in common. Their list might include cliffs, ridges, steep slopes, rocks, trees, structures, people, etc. Encourage students to define a landslide based on what they see in the pictures. Explain to students that landslides are rock, earth, or debris flows on slopes that move due to gravity.

(Optional) Explain gravity to your students. Gravity is the driving force behind landslide flow. Gravity is the attractive force between all massive objects. It causes apples to fall from trees towards the Earth, stars to pull planets into orbits, and cannonballs that are thrown skyward to return to the Earth. Gravity produces the weight of an object, which can cause an object to move down an inclined surface. The resisting forces are forces that cause the landslide material to resist the downward pull of gravity. Two primary resisting forces in this example are frictional forces (discussed already in Lesson 6) and forces applied by the weight of the material (stress).

In general, the landslide driving force is most heavily influenced by the weight of the potential landslide objects and the inclination angle. When an object or group of objects rests on a horizontal surface with an inclination angle of 0°, the pull of gravity produces no landslide driving force because all the weight is pushing perpendicularly into the surface. The higher the inclination angle, the more dominant gravity becomes in "pulling" material down the slope. This is because part of the weight starts to "pull" the object in the direction of travel along the surface (as depicted in blue and labeled "Landslide Driving Force"). Landslides begin to occur when the resisting forces reach a limit due to the strength of the material, the frictional properties between the slide material and the bedrock, or both. (The resisting forces are depicted in red and labeled "Landslide Resisting Forces"). When the landslide resisting forces are equal in magnitude and opposite in direction to the landslide driving force, the object or group of objects will not move. When the landslide driving force becomes greater than the landslide resisting forces, the object will begin to move, which is analogous to a landslide in this setup. Triggering events such as earthquakes, heavy rain, or upsetting the inclined surface through digging can help to initiate a landslide, but gravity is always the primary force that enables any landslide to occur, regardless of how that landslide was triggered.

Explain to the students that there are three phases of landslide behaviour: slope failure, transport of materials, and final deposition of slide materials. Ask them to look for where these phases take place in the photographs. Explain that slope failure can be gradual or sudden, and can exhibit noticeable properties such as visible cracks, slumps, or loud snapping sounds as slide materials break free from the bedrock or base material. Transport of materials can also be gradual or sudden, and can exhibit a wide range of transport modes such as rock or debris falling; rock or debris tumbling; material clump tumbling or sliding; fluid flow; or some combination of any of these modes.

Three Friends in a Valley

This Tabletop Exercise specifically relates to the landslide hazards associated with earthquakes; however, landslides describe many different types of motion of sand, soil, mud, rocks, and other earth structures. Landslides occur in response to a wide variety of natural and manmade triggering events, such as earthquakes, heavy rainfall, volcanic activity, or road and building construction. There are more variations in landslide mechanisms and triggering events than can be covered in an introductory course. Therefore, the most important outcome of any lesson on landslides is that students develop the curiosity and willingness to critically evaluate what their surroundings can tell them about landslides.

Read the following scenario while stopping to ask questions and discuss the material with your students at the indicated points or when students ask questions that are relevant to the discussion of earthquake hazards:

Three friends (Sara, Amira, and Gozen) live in the small city of Shahrabad, which is located in a beautiful mountain valley. The bottom of the valley has a small river running through it. The walls of the valley have land that includes forests and farms. The friends have lived there since they were young and they know that earthquakes sometimes happen there. They have only felt one small earthquake, but their parents and grandparents have told stories about some strong earthquakes that have happened in the area. Sometimes, during extreme weather like heavy snow or rain, the road that comes into Shahrabad from a nearby city is closed because rocks have fallen on the road or the road has washed away.

Sara and Amira live next to each other on farms located on slopes in the valley. Sara's farm used to have a natural spring at a crack between two rocks that produced drinking water for both Sara's and Amira's families, but the spring stopped producing water about a year ago. Recently, a neighbour has started complaining that some parts of his land have become very soggy and soaked with water, especially near the bottom of the valley.

What are Natural Springs, and What are a Couple of Reasons Why the Spring on Sara's Farm Stopped Giving Water?

Potential answers: Springs occur when water flows through cracks below the Earth's surface. The water can be a mix of rain water, water from underground channels that travel downhill towards the river, or water that is pushed up from deep underground in the deepest parts of the Earth, which has not ever before been to the surface. Sometimes springs located very close to each other on the surface of the Earth have completely different paths that the water in each follows. The water that is soaking the neighbour's land may or may not be related to the water that used to come out of the spring; however, both of the changing events indicate that the land that Sara, Amira, and the neighbour live on is undergoing movements that may not be visible on the surface.

The spring might have stopped because of some small change in the path of the water due to small movements of the ground, or because the source of the water has become empty. The changes in the path of the water could have occurred deep in the Earth or just a couple of metres beneath where the spring is located. When the water flows through narrow cracks, very small shifts in the ground can stop the flow of water.

Sara's and Amira's farm share a wooden fence to keep their farm animals from wandering around. Sara and Amira often climb over the fence to play in the forest around their farm. About three years ago, they noticed that the fence posts were sloped at an angle at one spot in the fence near their path to the forest, and they were concerned that climbing over the fence was pushing the fence over. They changed their path so they didn't have to climb over the fence and then gradually forgot about the sloping fence posts. But the fence posts continued to tip over, little by little, without anyone noticing the low part of the fence. Until one day, about a month ago, a donkey got away by jumping over the low part of the fence. They helped their fathers fix the fence and straighten the fence posts so the donkey couldn't get away.

What are Some Possible Reasons for Why the Fence is Slowly Tipping over?

There are many answers possible that don't relate to landslide hazards. The fence could be old and the wood falling apart. The donkey could be pushing on the fence to eat some tasty grass that grows outside of that part of the fence. But the ground could also be moving very slowly beneath the farm, causing the fence posts to point uphill over the years. The fact that the spring stopped giving water may support this idea even further, especially if the path of the water to the surface was broken because the ground had shifted very slightly.

Gozen lives down in the city in a house. Sometimes all of the friends gather there to have dinner and listen to the radio or watch television. From where her family eats dinner, they can see the river. Her father helps to build and fix pipes that move water for farmers in the valley, and he also helps to build and fix houses. A wealthy man has just built a house above a very steep hill that has a beautiful view of the valley, and he even paid just to have electricity from the city strung on wires up the hill to the house. But the rooms already have cracks in the walls on the side of the house near the steep hill. Some of the windows and doors have also become very difficult to open and close. Gozen's father has been working there the past few days and he jokes about how the wealthy man complains that his house was not built very well by workers from a nearby city.

What are Some Possible Reasons for the Cracks in the Walls? What are Some Ways to Find out What is Really Happening?

Again, the wealthy man may be right and the walls were indeed poorly built. Oftentimes, houses also settle naturally as they age and cracks form as the house comes to rest on the ground. However, the cracks are forming on the walls on the sides nearest the steep hill, which may indicate that the part of the house that rests on ground above the steep hill may be on unstable ground that is slowly creeping down the hill. Doors and windows can become difficult to open and close because the house is changing shape as the ground moves beneath it, causing the frames to become misshapen. Also, if the ground was naturally unstable prior to building the house, the added load of the new house may be speeding the rate of movement of the creeping slope. Unstable ground or ground that is creeping is much more likely to release during a triggering event such as an earthquake or heavy rainfall. There are many ways to tell what the real cause of the cracks may be. Other indications, such as the bending of pipes, fences, footpaths, or roads, can be found to see if the ground is moving. If the ground is shifting, then electrical wires attached to polls in the ground near the edge of the hill will become very tight as the polls move with the ground.

One day, the three friends decide to go play in the forest together. They travel farther up the hill than they had ever gone before. They find a very interesting bunch of very tall trees whose trunks grow out of the ground at an angle before the trees turn straight and point up into the air like a normal tree. Some of the trees have such a sharp angle that the girls can sit in the angle of the trees like a comfortable chair with their feet dangling down the slope of the hill! Most of the trees are curved in the same direction in the middle. The three friends name it the Sideways Forest.

What would Cause Trees to Grow Like This?

Figure: *A common tree shape formed due to ground creep*

Trees always grow up towards sunlight, so presumably the trees initially grew at a different angle when they were young. The fact that the trees were all curved in the same direction, and that they were all located next to each other, might indicate that the ground beneath the Sideways Forest is all shifting in one direction. The trees are all much older than the girls, implying that the ground has been moving for a very long time. This might mean that the ground above the farm is unstable and could be dislodged in the event of heavy rain, an earthquake, or human activities like road construction.

One day, while the friends are walking back home from school, there is an earthquake. It is strong enough to shake many of the buildings around them, and the earthquake is over after about a minute. They are just as far away from Gozen's home as from Sara's and Amira's farms.

Where should the Friends Go First?

There are many reasons to go to Gozen's house first. Gozen has a radio and television, so they can hear about the damage caused by the earthquake and whether emergency services are being delivered. The radio or television, if they are functioning immediately after the earthquake, may also have information on any developing weather system that may be coming in that could make the situation created by the earthquake even worse, such as heavy rain or snowfall. In addition, the combination of observations the girls have noticed around Amira's and Sara's farms indicate that the ground might be unstable and prone to landslide if another earthquake occurs.

Knowing that the farmland is unstable, it is natural for the girls to want to make sure that their families and homes are safe. At that moment it is very dangerous to go there because the possibility for aftershocks is high. Since the girls are safe, they should first make contact with a parent or family friend to let their parents know they are safe and find out what has happened in order that they can make an informed decision about what to do next, while conserving water, food, and medical supplies.

The families are all fine, and they meet at Gozen's house to talk about what happened. Through the radio they find out that there has been an earthquake that has caused numerous landslides throughout the region. The neighbour whose land was becoming soaked with water reported that, in some places on his land, the surface had broken into cracks and the smooth slope had become shaped like stairs. The road has been blocked by some falling rocks, but the families have some food stored away for when the road is closed. Gozen's dad says that many pipes have been broken in several places, so there is no water to be gathered through the city's water system. They send the friends down to the river to gather some water to support the families. While the three friends are at the river, they notice that the water level is much lower than it had been the day before.

What are Some Possible Causes for the Low River Water Level, and What should the Girls Do about it?

In river valleys that are likely to experience landslides after earthquakes, a sudden decrease in river water levels may indicate a landslide dam has formed upstream of the city. A landslide dam occurs when a landslide has blocked a river or stream, causing water to build up behind it. This causes flooding upstream and a drought or decreased water flow downstream. Landslide dams can be extremely dangerous because they are usually highly unstable. As the water builds up behind the dam, the landslide becomes saturated with water and can break catastrophically, flooding all areas downstream with little or no warning. Recall the instability of water-saturated unconsolidated materials observed during the liquefaction exercise.

The three friends should notify their parents or other city officials immediately of this possibility so that they can determine whether a landslide dam has formed. If action is taken quickly, the water behind the landslide dam can be released gradually before it builds up to dangerous levels. Even children can save entire communities!

The three friends told their parents immediately about the water level, who in turn alerted city officials. A small landslide dam had formed upriver, but it was not large enough to be a concern. All three families stayed at Gozen's house for a few days as aftershocks were felt, but none of them was as big as the original earthquake. While there had been no landslides occurring on their farms during this series of earthquakes, the families became concerned about future earthquakes or other triggering events that could cause them to lose their farmlands and houses. They began to discuss ways of preventing landslides from taking place on their land.

What are Some of the Things the Families can Do to Prepare for Landslides, and to Prevent Landslides from Occurring on Their Land?

Recognizing and communicating the signs of unstable land to city officials and neighbours in the community as well as preventing further human development that can cause a landslide or put more lives in danger (e.g., building more human structures on unstable land) are the first steps to mitigating future hazards.

Ask the students to make a list of whom they should contact if they suspect there is an emergency involving landslides. The game would be to rank the list in order of those the students should contact first after an event or warning sign has occurred. Time and resources permitting, this ranking could be organized into a letter, sent to a local emergency services official, and evaluated for technical proficiency. The local emergency services official should be able to provide guidance about additional resources in the community that were not identified as well as how students' responses and actions in an emergency can help or hinder the planned relief activities.

The second step is to prepare for future events by establishing safe, emergency meeting places in your community. People should also store food, water, blankets, and medical supplies in case emergency services are unable to reach them for many days following an earthquake or landslide. The students will be taught how to prepare for emergencies in upcoming lessons.

The most difficult steps involve how to stabilize land that already shows signs of instability. Planting trees on otherwise open, exposed slopes may help to reduce water within soil, which can trigger landslides more easily. Deep root systems of trees may help to bind loose soil and debris that would otherwise be easily released during a landslide. There are other, more costly engineering steps, such as creating landslide barriers and ensuring proper drainage across roads, that can be undertaken at the community level to mitigate landslide hazards. Whenever possible, these steps should be coordinated with other community members to help reduce the costs and ensure that everyone is aware of the risk of landslides.

Exercise Summary and Post-Exercise Discussion

Figure: *Many warning signs of unstable landslide-prone areas may be visible at the surface of the Earth. Now that students have completed the exercise, ask them to brainstorm about what evidence they would look for to indicate landslide-prone areas. Have them compare and contrast what they knew or did not know before the exercise started with what they learned as the exercise progressed. Emphasize that the point of this exercise was to put the students into a situation without any clear answers provided before the situation began, which is closer to how a real emergency situation unfolds.*

Steep slopes, old landslide scars, and new cracks at the surface are the most obvious signs of the potential for landslides. Other subtle indications may include: displaced fences, roads, or overhead power lines; leaning poles driven into the ground; new springs or water seepages in places that have never really been wet before; sticking windows and doors in houses due to shifting house frames; and unusual increases or decreases in creek water levels that cannot be attributed to rainfall or weather patterns.

Experiment: The Homemade Landslide

Now that the students have simulated some common landslide experiences, they will have the opportunity to create, study, and describe a hands-on landslide demonstration. This simple experiment will permit students to observe the three phases of landslide development (slope failure, transport of materials, and final deposition of slide materials) and explore the differences between the experiment and actual landslides.

Materials

- Metal cookie or bread pans with 1–5 cm tall edges, 40–60 cm in length and width
- Pencils
- Protractors
- Sand
- Soil
- Gravel
- Flat rocks
- Cloths or paper towels
- Newspapers to cover the working surfaces
- Water pitcher with water (optional)
- Toothpicks (optional)

This activity can be done either as an entire class with one pan or in groups with many pans and enough supervision. However, this activity can easily become messy, so it is not recommended that the class be split into more than 2–3 groups. The behaviour of the sand, soil, gravel, etc. will depend on the particular properties of the materials that have been gathered. It is highly recommended that at least one educator go through all the activities once before conducting the class in order to identify any problems with the setup as well as to figure out methods of minimizing the mess created during certain parts of this activity. It may be necessary to move the activity outdoors, or to conduct all the tilting experiments inside a large tray to catch water and spilling material.

Procedures

Cover the area that you intend to use for the demonstration with newspapers. Create a data table to record the results ofyour experiment runs. Encourage students to add new columns, to experiment with other materials that are not mentioned above, or to make remarks about the behaviour of the system that does not fit neatly into any of the categories.

Assigning duties: assign one student to be in charge of lifting one side of the cookie pan to create an incline, telling the student to lift it slowly and smoothly, without shaking the pan or stopping, until the material on the pan begins to slide down. Assign another student to be in charge of measuring the angle that the cookie pan is at using the protractor. This student should call out the angle in increments of 5° when the material is not moving, and should call out the angle when the material moves For instance, the material does not move until it reaches 23° where it begins to crack, then at 27° it collapses. The student would read out, –5 degrees... 10 degrees... 15 degrees... 20 degrees . . . cracking at 23 degrees [as cracking begins], 25 degrees, collapse at 27 degrees." This student should be concentrating on reading out the angle when significant events occur; meanwhile, the remaining students should be making specific notes that describe what is happening.

Start the experiment with the dry materials first, in order of those that are easiest to clean up (example: flat rocks, then gravel, then sand, and then soil). Have the students complete the entries in the table that describe how the landslide simulation occurred. Have the students record details of the experimental setup before tilting the pan (e.g., what material was used, the dimensions ofthe pan, how deep the material sat in the pan). As the material begins to move when the pan is tilted, encourage the students to describe and note what happens in the data table at different angles, as provided in tables. After the materials have collapsed, have the students make note of how the material collapsed: did the material flow like a fluid, or did it travel in discrete blocks or slabs; did it come to rest as a smooth pile of material, or were there small bumps or cracks in the collapsed material?

Complete the first run with all the materials and discuss. If time permits, do more than one experiment for each material. When changing materials, use the cloths or paper towels to wipe off the cookie pan in order to make the conditions of the different experiments as similar as possible.

The key to any scientific experiment is reproducibility. If the surface of the cookie pan is not regularly cleaned between different experiments, then the results might vary because of debris left on the surface from the previous experiments.

Discuss what happened with the students and encourage them to derive new questions about the setup. Were there any patterns? Which materials collapsed at the smallest and largest angles? Which materials collapsed quickly and which collapsed slowly? Were the materials used similar to materials that make up the mountains or hills around their region; why or why not? Could the students identify movements or behaviours in the materials that collapsed which corresponded to the warning signs that the three friends observed around their city in the Tabletop Exercise scenario?

Ask the students to predict what will happen if they repeat the experiments again with the same materials, except that this time the pan will be tapped or gently shaken while it is being titled. Repeat the steps above exactly as before for all the same materials, but this time ask the student assigned to tilt the cookie pan to gently shake or tap the cookie pan as it is being lifted. Instruct the student to make sure that the rate at which the cookie pan is lifted is about the same as before. Instruct the student to keep the shaking motion about the same (in magnitude of motion and frequency) while tilting for all materials. Record any differences the students observe about the angles at which the materials collapse, how fast they collapse, the ways in which the materials collapsed, etc. The overall result should be that the angle of collapse should be lower when the pan is shaken. Ask the students to describe why this should be the case, based on their knowledge of the weight, direction of gravity, friction, and degree of inclination angle. One answer is that materials like sand, when viewed very closely, consist of tiny little specks — each tiny stone resting on many other tiny stones and surrounded by pockets of air called pore spaces (recall the liquefaction exercise in Lesson 7). When the students shake the cookie pan with pebbles on the slope, they tend to either roll down the slope or be pushed out directly away from the face of the incline, which causes it to "jump" down the slope. When you add up the contributions from countless little stones that are all bouncing off one another, the net effect is that the entire mass of material is pushing itself down the slope as every little stone hops a tiny distance down the decline.

Finally, for each material, select an angle lying directly between the respective angles of collapse for the shaking and non-shaking cases (example: the sand collapses at 15° when shaking and 25° when not shaking, so you should select the angle of 20° for this case). Prepare each material in the pan and lift to the selected angle for that material. Ask the students what they believe will happen if the pan is shaken and why. The answer should be that it will collapse because we are moving the material from a stable to an unstable case by changing its environment. This is analogous to an earthquake-induced landslide. Have the student shake the pan exactly as before, and have the students note the results.

The actual results may vary, but this is also part of the exercise. In some cases, the material will collapse as predicted; in others, it may not. There is a tremendous variety of factors that dictate when, how, and why a landslide will occur, such as how compact the material sitting in the pan is, random variations in shaking the pan as it is tilted, and the random arrangement of sand grains at the contacts with the cookie pan, to name a few.

Summarize all of the observations made. Encourage the students to discuss the limitations of this setup when relating it to a natural environment, and have them identify some of the challenges Earth scientists may face when trying to prevent landslides from affecting human communities. There are many possible answers: natural materials are not homogeneous but are a mix of different materials and consistencies. Natural slopes are uneven and irregular with complex surfaces, cracks, and forces. The presence of water is also a complicating factor: for some materials observed with this type of setup, the water may increase cohesion and raise the angle of collapse. But in the real world, too much water tends to lower the angle of collapse due to the increase in pore pressure.

Encourage students to use their imaginations and the materials at hand with the sand, gravel, and soil materials to stabilize the surface. Some ideas include simply placing a wet paper towel over the surface. This functions very similar to surface netting and is a very effective, but somewhat expensive, way of stabilizing real surface cuts next to roads and buildings. Other ideas include building retaining walls using the flat rocks or pebbles from the gravel to hold back the material. Have the students experiment with different stacking arrangements and shapes at the base of the cookie pan as they tilt it upward to see who can build the structure that best prevents landslides. Encourage students to find other materials around the classroom (e.g., sticks, string, pencils) that may be incorporated into landslide barriers.

(Optional) Time permitting, have the students experiment with wetted materials. Have the students completely soak some of the materials in water for a few minutes (the longer the better). Have them load the cookie pan as before and note the results of raising the cookie pan. The results should be highly unpredictable. In some cases, the material will collapse at a shallower angle. In others, the material will collapse at a higher angle, but the result will be much more catastrophic and sudden. The students can simulate rain water infiltration by very slowly pouring water onto the material at the top of the cookie pan as it is lifted. The students can also experiment with inducing an earthquake at various angles with wet materials.

Other optional activities include simulating the surface effects of creep by placing toothpicks (driven vertically into the sand, soil, or gravel materials) and slowly lifting up the cookie sheet as in previous steps. For some materials, very small movements of the material are reflected in the steady movement of the toothpicks before the slope collapses.

4

Mass Wasting

Mass wasting, also known as slope movement or mass movement, is the geomorphic process by which soil, sand, regolith, and rock move downslope typically as a mass, largely under the force of gravity, but frequently affected by water and water content as in submarine environments and mudslides. Types of mass wasting include creep, slides, flows, topples, and falls, each with its own characteristic features, and taking place over timescales from seconds to years. Mass wasting occurs on both terrestrial and submarine slopes, and has been observed on Earth, Mars, Venus, and Jupiter's moon Io.

When the gravitational force acting on a slope exceeds its resisting force, slope failure (mass wasting) occurs. The slope material's strength and cohesion and the amount of internal friction between material help maintain the slope's stability and are known collectively as the slope's shear strength. The steepest angle that a cohesionless slope can maintain without losing its stability is known as its angle of repose. When a slope possesses this angle, its shear strength perfectly counterbalances the force of gravity acting upon it.

Mass wasting may occur at a very slow rate, particularly in areas that are very dry or those areas that receive sufficient rainfall such that vegetation has stabilized the surface. It may also occur at very high speed, such as in rock slides or landslides, with disastrous consequences, both immediate and delayed, e.g., resulting from the formation of landslide dams.

Factors that change the potential of mass wasting include: change in slope angle, weakening of material by weathering, increased water content; changes in vegetation cover, and overloading.

The Importance of Water in Mass Wasting

Water can increase or decrease the stability of a slope depending on the amount present. Small amounts of water can strengthen soils because the surface tension of water increases soil cohesion. This allows the soil to resist erosion better than if it were dry. If too much water is present the water may act to increase the pore pressure, reducing friction, and accelerating the erosion process and resulting in different types of mass wasting (i.e. mudflows, landslides, etc.). A good example of this is to think of a sand castle. Water must be mixed with sand in order for the castle to keep its shape. If too much water is added the sand washes away, if not enough water is added the sand falls and cannot keep its shape. Water also increases the mass of the soil, this is important because an increase in mass means that there will be an increase in velocity if mass wasting is triggered.

Types of Mass Movement

Types of mass movement are distinguished based on how the soil, regolith or rock moves downslope as a whole.

Creeps

Soil creep is a long term process. The combination of small movements of soil or rock in different directions over time are directed by gravity gradually downslope. The steeper the slope, the faster the creep. The creep makes trees and shrubs curve to maintain their perpendicularity, and they can trigger landslides if they lose their root footing. The surface soil can migrate under the influence of cycles of freezing and thawing, or hot and cold temperatures, inching its way towards the bottom of the slope forming terracettes. This happens at a rate that is not noticeable to the naked eye.

Landslides

A landslide, also called a landslip, is a rapid movement of a large mass of earth and rocks down a hill or a mountainside. Little or no flowage of the materials occurs on a given slope until heavy rain and resultant lubrication by the same rainwater facilitate the movement of the materials, causing a landslide to occur. The common forms of landslides are slump, debris slide, rock slide, rock fall, debris fall and avalanche.

Flows

Movement of soil and regolith that more resembles fluid behaviour is called a flow. These include avalanches, mudflows, debris flows, earth flow, lahars and sturzstroms. Water, air and ice are often involved in enabling fluidlike motion of the material.

Topples

Topples are instances when blocks of rock pivot and fall away from a slope.

Slump

A slipping of coherent rock material along the curved surface of a decline. Slumps involve a mass of soil or other material sliding along a curved surface (shaped like a spoon). It forms a small, crescent-shaped cliff, or abrupt scarp at the top end of the slope. There can be more than one scarp down the slope.

Falls

A fall, including rockfall, is where regolith cascades down a slope, but is not of sufficient volume or viscosity to behave as a flow. Falls are promoted in rocks which are characterised by presence of vertical cracks. Falls are a result of undercutting of water as well as undercutting of waves. They usually occur at very steep slopes such as a cliff face. The rock material may be loosened by earthquakes, rain, plant-root wedging, expanding ice, among other things. The accumulation of rock material that has fallen and resides at the base of the structure is known as talus.

Triggers of Mass Wasting

Soil and regolith remain on a hillslope only while the gravitational forces are unable to overcome the frictional forces keeping the material in place. Factors that reduce the frictional resistance relative to the downslope forces, and thus initiate slope movement, can include:

- seismic shaking
- increased overburden from structures
- increased soil moisture
- reduction of roots holding the soil to bedrock
- undercutting of the slope by excavation or erosion
- weathering by frost heave
- bioturbation

Submarine Landslides: Processes, Triggers and Hazard Prediction

Huge landslides, mobilizing hundreds to thousands of km^3 of sediment and rock are ubiquitous in submarine settings ranging from the steepest volcanic island slopes to the gentlest muddy slopes of submarine deltas. Here, we summarize current knowledge of such landslides and the problems of assessing their hazard potential. The major hazards related to submarine landslides include destruction of seabed infrastructure, collapse of coastal areas into the sea and landslide-generated tsunamis. Most submarine slopes are inherently stable. Elevated pore pressures (leading to decreased frictional resistance to sliding) and specific weak layers within stratified sequences appear to be the key factors influencing landslide occurrence. Elevated pore pressures can result from normal depositional processes or from transient processes such as earthquake shaking; historical evidence suggests that the majority of large submarine landslides are triggered by earthquakes. Because of their tsunamigenic potential, ocean-island flank collapses and rockslides in fjords have been identified as the most dangerous of all landslide related hazards. Published models of ocean-island landslides mainly examine 'worst-case scenarios' that have a low probability of occurrence. Areas prone to submarine landsliding are relatively easy to identify, but we are still some way from being able to forecast individual events with precision. Monitoring of critical areas where landslides might be imminent and modelling landslide consequences so that appropriate mitigation strategies can be developed would appear to be areas where advances on current practice are possible.

Submarine landslides are one of the main agents through which sediments are transferred across the continental slope to the deep ocean. Sediments derived from land (mainly carried by rivers) and from the continental shelf (e.g. through erosion and transport by ocean currents and storms) are, in the first instance, often deposited on the upper continental slope where they may be interbedded with pelagic sediments settling from the water column. However, these deposits often equate only to temporary storage, as instability in the slope deposits leads to periodic slope failure, landsliding and onward downslope transport. The human consequences can be severe and include destruction of seabed infrastructure (e.g. telecommunications cables), subsidence of coastal areas and generation of damaging tsunamis. In this latter respect, flank collapses on oceanic islands, which consist largely of rock (rather than sediment) and have both subaerial and submarine components, pose a particular threat. The aim of this paper is to summarize key aspects of submarine landslides, especially where these relate to their hazard potential, and assess our ability (or lack of ability) to predict where and when future submarine landslides will occur.

Landslide terminology is beyond the remit of this paper, but as noted by several authors (Hampton *et al.* 1996; Mulder & Alexander 2001; Canals *et al.* 2004) the reader should be aware of the complexity of landslide nomenclature and the frequent imprecise use of landslide terminology, especially in the submarine environment where information on landslide processes is often limited. In this paper, 'landslide' is used as a generic term encompassing all forms of slope failure, irrespective of process. Other terms used in this paper, including 'slide', 'debris flow', 'debris avalanche' and 'turbidity current', each imply a particular process as defined below:

- Slide: movement of a coherent mass of sediment bounded by distinct failure planes.
- Debris flow: laminar, cohesive flow of clasts in a fine-grained matrix (e.g. wet concrete).
- Debris avalanche: rapid flow of cohesionless rock fragments with energy dissipation by grain contact.
- Turbidity current: gravity flow in which sediment grains are maintained in suspension by fluid turbulence.

Occurrence, Distribution and Scale of Submarine Landslides

Landslides are widespread on submarine slopes, particularly in areas where fine-grained sediments predominate. In the North Atlantic, this corresponds mainly to areas at high and low latitudes (Weaver *et al.* 2000) and appears to correlate with the weathering style of rocks on land. In general, glacial action at high latitudes and chemical weathering processes at low latitudes produce fine-grained sediments that form thick accumulations on the continental slope and appear to favour landslide formation. Between these zones, at mid-latitudes, fluvial weathering and sediment transport produces greater quantities of coarser grained sediment. This sediment is often transported directly to the deep ocean basins by turbidity currents, which pass through submarine canyons and bypass the slope. Although small landslides frequently occur on canyon margins, adjacent slopes are relatively starved of sediment and appear less likely to be affected by landslides. The submarine deltas and fans of large rivers are also subject to widespread landsliding, related to the rapid accumulation of fine-grained sediments on the continental slope (e.g. Mississippi Fan; Prior & Coleman 1982). Landslides in fjords (Jorstad 1968) and on the flanks of oceanic islands pose particular hazards to humans. In fjord environments, landslides are frequently associated with 'quick clay', a particularly unstable sediment created when marine clays are uplifted above sea level (usually by glacial rebound) and leached by fresh water (Locat *et al.* 2003). Failure of submarine deltas formed where rivers discharge sediment onto the steep submarine walls of fjords is also common (Prior *et al.* 1982; Kulikov *et al.* 1996). Landslides on ocean-island flanks have been cited as posing a tsunami threat on a trans-oceanic scale (Ward & Day 2001), potentially comparable in magnitude to or even larger than the 2004 Indian Ocean tsunami (Lay *et al.* 2005). While this extreme scenario can be questioned (Mader 2001; Wynn & Masson 2003), there is little doubt that the threat posed by ocean-island flank landslides is both real and significant.

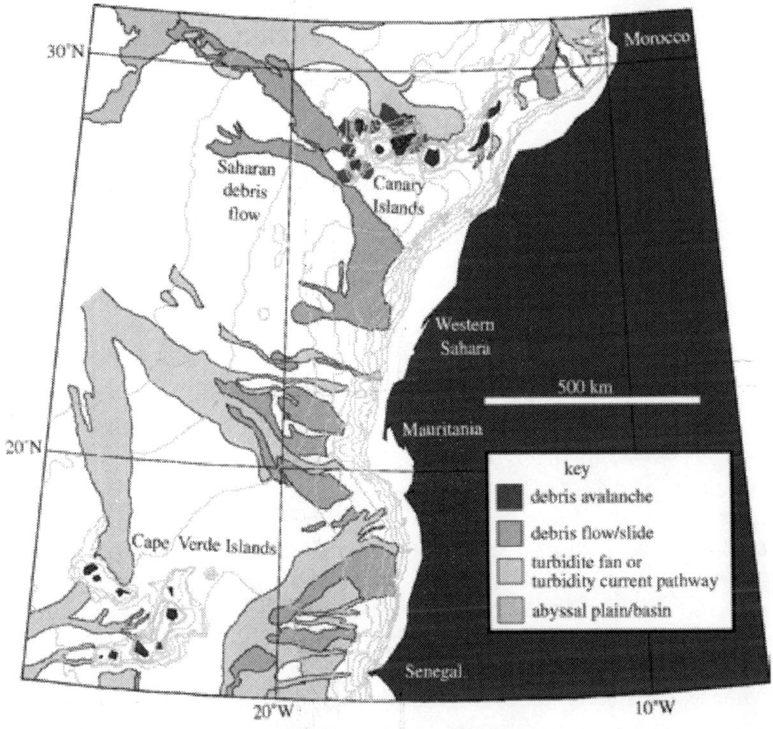

Figure: *Distribution of landslides and landslide deposits on the seafloor of the NW African margin. The map shows the degree to which this margin has been impacted by landslides in the recent geological past. When extrapolated into three dimensions the importance of landslides in the margin building process is obvious.*

Hühnerbach *et al.* (2004) have recently documented the distribution of landslides on North Atlantic margins. Although the quality and completeness of many of the published descriptions of submarine landslides were limited, these authors were able to conclude that:

The occurrence of landslides is not greatly influenced by slope gradients, except at the shallowest gradients where relatively few landslides occur.

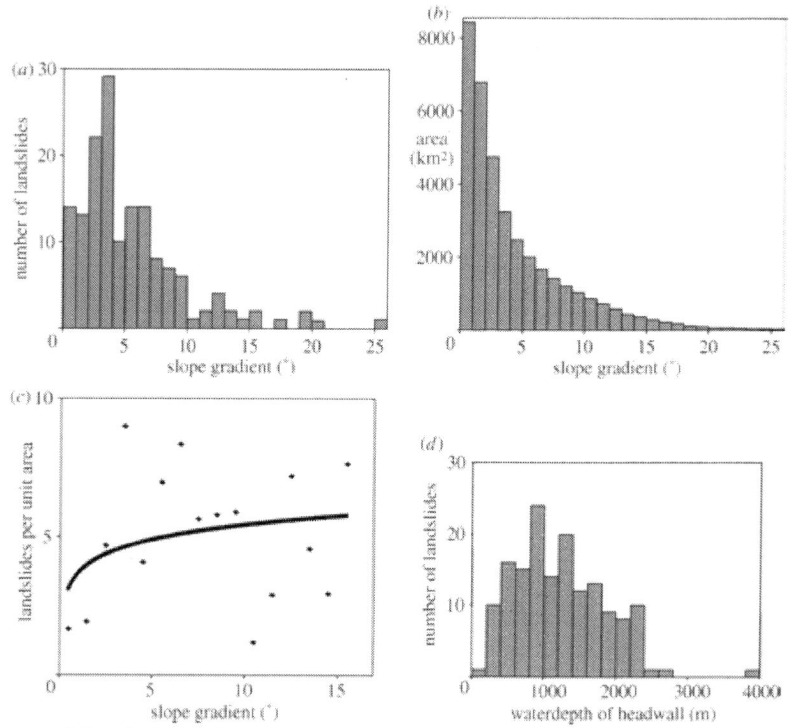

Figure: *Numbers of landslides on the eastern continental slope of North America between 30 and 45°N, plotted against slope in the landslide headwall area.* (b) *Distribution of slope gradients between 36 and 37°N and between 200 and 3000 m water depth on the eastern North American slope (extracted from multibeam data gridded at 100 m). Comparison of the histogram shapes in* (a) *and* (b) *suggests that landslide occurrence is largely independent of slope angle at gradients more than 3°, but that landslides are relatively rarer on lower slope gradients. A plot of landslide occurrence per unit area against slope gradient* (c) *gives some support to the qualitative comparison of* (a) *and* (b) *based on histogram shape, although the correlation in* (c) *is weakened by the scatter of data points which is mainly due to the low numbers of observed landslides.* (d) *Number of landslides plotted against water depth in the headwall region, indicating that most landslides are initiated in mid-slope. Similar distributions of landslides are seen in the eastern Atlantic confirming the robustness of the interpretations based on* (a)*–*(d) *above (Hühnerbach* et al. *2004).*

There is a suggestion that the largest landslides on the continental slope occur on the lowest slopes, often as low as 1°.

The greatest number of landslide headwalls occur on the mid-slope, with a peak at 1000–1300 m waterdepth, rather than at the shelf edge or on the upper slope as might be expected. The largest submarine landslides can involve many thousand km^3 of material, two to three orders of magnitude larger than any terrestrial landslide. For example, the Storegga slide involved some 3000 km^3 of sediment, affected 95 000 km^2 of the Norwegian slope and basin and had a run-out distance of around 800 km (Haflidason *et al.* 2004). To put this into perspective, the area is about 20% bigger than Scotland and the run-out is close to the length of mainland Britain. The largest landslides occur mainly in two settings, on open continental margins and on oceanic island flanks. This appears to be a function of specific aspects of the geology and morphology of these areas. Continental margin slopes that are subject to large-scale failures are typically of low gradient (from less than 1° to 5°) with gentle topography; however, the 'drop' from shelf edge to basin floor can be up to 5 km over distances of a few hundred kilometres. Parallel-bedded sediment sequences with little variability over large areas characterize their sub-surface structure, with the result that, should the conditions for slope failure occur, they can simultaneously affect large areas.

When measured from the top of the highest volcano to the bottom of the adjacent ocean basin, ocean islands such as Hawaii and the Canary islands have the greatest relief of any topographic feature on Earth. The island slopes can be very steep (e.g. on Tenerife in the Canary islands the average slope from the top of Teide volcano to the coast is 15°) and volcanic processes tend to build, load and steepen these slopes with time. Despite this, not all these slopes are unstable, with landslide occurrence apparently closely controlled by geology, particularly the trends of dyke intrusion (rift zones) on the islands (Moore *et al.* 1989; Carracedo 1996; Masson *et al.* 2002).

However, as discussed in the following section, a landslide's size is not necessarily proportional to the hazard it poses. In particular, continental margin landslides that occur on very low slopes, far from land, may form relatively slowly by retrogressive processes, similar to quick clay flows on land (Bentley & Smalley 1984). Many such landslides appear to have limited tsunamigenic potential, although they may still pose a threat to cables or other seabed installations.

Submarine Landslides and Hazards

The hazard posed by submarine landslides will vary according to landslide scale, location, type and process. Even small submarine landslides can be dangerous when they occur in coastal areas. The

1996 Finneidfjord slide, in northern Norway, mobilized only 0.001 km^3 of sediment and little of this material travelled more than a few hundred metres from source, but four people were killed when a house and car were carried away (Longva *et al.* 2003). The 1979 Nice airport slide also cut back onto land, killing several men working on the airport extension (Assier-Rzadkiewicz *et al.* 2000). However, the effects of this landslide were felt at least 100 km offshore, where a turbidity current generated by the landslide broke submarine telephone cables. A local tsunami, which resulted in the death of one person, was also observed. The 1929 Grand Banks earthquake resulted in submarine landslides, a turbidity current, and a tsunami that caused significant casualties (Heezen & Ewing 1952; Piper *et al.* 1999; Fine *et al.* 2005). This is one of the best-known submarine landslides because the resultant turbidity current broke several submarine cables sequentially downslope, allowing the speed of such a current (up to 30 m s^{-1}) to be measured for the first time. It also illustrates one of the main difficulties in submarine geohazard study—when a coupled earthquake/landslide generates a tsunami, which of the two hazards produces the tsunami or could their effects even be combined (Fryer *et al.* 2004; Fine *et al.* 2005)? The latter has been suggested for the 1998 tsunami that struck Papua New Guinea (PNG), killing over 2000 people (Satake & Tanioka 2003). However, there is now considerable evidence that many 'unusual' tsunamis, particularly those with high near-field run-ups that decay rapidly away from source, are directly caused by landslides (Bardet *et al.* 2003; Okal & Synolakis 2004). Rotational slides (often referred to as slumps), where a thick slide block with a steep headwall can move rapidly downward, may be particularly effective in generating tsunamis, even when the lateral distance moved is small and little effect is seen on the seafloor downslope of the immediate landslide site. The PNG tsunami is most likely to have been generated in this way (Matsumoto & Tappin 2003; Sweet & Silver 2003).

Finally, it should not be forgotten that an increasing proportion of the world's oil and gas is now recovered from deep-water areas offshore, where slope instability can be a major geohazard. The juxtaposition of the Ormen Lange gas field, which is set to supply some 20% of UK natural gas requirements in future years, and the Storegga slide, brings this sharply into focus.

Causes of Landslides

Many factors have been suggested as probable or possible contributors to the initiation of submarine landslides. These vary from sudden impacts operating on timescales of minutes (e.g. shaking due to earthquakes) to geological processes operating on timescales of tens to hundreds of thousands of years (e.g. climate change; Weaver & Kuijpers 1983). Broadly, they can be divided into two types, those related to the geological characteristics of the landslide material (e.g. overpressure due to rapid deposition or the presence of a weak layer) and those driven by transient external events (e.g. earthquakes or climate change). In some cases, the relationship between a landslide and its cause is very obvious and direct (e.g. failure of an oversteepened slope). In others this relationship may not be obvious at all, especially where the link is indirect and involves some intermediate process. A good example of this is found on the Norwegian continental slope where the locations of landslides, such as the Storegga and Traenadjupet slides, can be related to the occurrence of specific geological horizons that act as weak layers (Laberg *et al.* 2003; Kvalstad *et al.* 2005). The distribution of these weak layers, however, is controlled by regional changes in sedimentation style, which, on an even broader scale, are driven by climate change between glacial and interglacial conditions (Bryn *et al.* 2005). On top of all of this, it has been calculated that an earthquake was still required to ultimately trigger these landslides (Bryn *et al.* 2005; Kvalstad *et al.* 2005)!

The above example shows that many if not all landslides result from the coming together of more than one of the factors given in table presented. Most continental slope sediments, for example, are stable under all but the most extreme earthquake shaking, unless other factors, such as overpressured layers or sensitive sediments are present (Roberts & Cramp 1996; Kvalstad *et al.* 2005). Many large historical landslides, recognized because they broke telephone cables or caused collapse of river deltas in fjords or tsunamis, have coincided with earthquakes (Tappin *et al.* 2001; Fryer *et al.* 2004; Fine *et al.* 2005). This presumably reflects the long term persistence of 'geological' factors compared to the transient nature of earthquake shaking, such that the short-lived transient event is the final push which tips the balance between a stable slope and a landslide.

Human activities can also contribute to submarine landslide triggering. Although landslides generated in this way are usually small, they can often be very dangerous because they extend upslope onto land and can affect inhabited areas. Examples include the 1996 Finneidfjord landslide, where it has been hypothesized that the trigger was heavy rainfall combined with rock blasting (Longva *et al.* 1996), and the 1979 Nice Airport landslide, where loading of the upper slope during construction work at the airport probably contributed (Assier-Rzadkiewicz *et al.* 2000). In other areas where submarine landslides have directly affected human-made structures, such as where buildings on deltas have been submerged or offshore oil installations destroyed by landslides, it is often impossible to determine (and sometimes unlikely) that human activities triggered the landslides. However, even if human activity is not to blame for triggering such landslides, building in these areas has undoubtedly contributed to the landslide consequences.

The concept that 'weak layers' oriented parallel to sedimentary bedding might control the location of many continental slope landslides is not new. However, it required the advent of modern seafloor survey technology, such as swath bathymetric mapping and 3D seismic systems, to demonstrate that this concept was correct. Indeed, it seems that submarine landslides at all scales are often controlled in this way (Lastras *et al.* 2004; Wilson *et al.* 2004; Bryn *et al.* 2005). However, we know very little about the nature and characteristics of these weak layers, since they have rarely been sampled and very little geotechnical work has been done on sediments recovered from them. The weak layers that underlie parts of the Storegga slide are a notable exception (Bryn *et al.* 2005; Kvalstad *et al.* 2005). Here it has been shown that the weak layers are composed of contourites (sediments deposited by ocean currents flowing along the continental slope) that are clay-rich and have higher water content and greater plasticity than the overlying less-well sorted glacial and glaciomarine sediments. As a result, the contourites are more sensitive and brittle (i.e. they lose strength rapidly when their strain bearing capacity is exceeded). Rapid loading of the water-rich contourites by glacial sediments appears to have raised pore pressures within the contourites and is the main factor contributing to landsliding.

Many sedimented slopes prone to submarine landslides show a history of landsliding that extends back through geological time. This observation can often be applied at quite local scales, with areas showing stacked landslide deposits sharply demarcated from those showing long-term stability (Solheim *et al.* 2005). The same is true of some volcanic island slopes, for example in the Canary islands, where part of the north flank of Tenerife has experienced at least five landslides in the last 2–3 Myr, while adjacent regions have seen none (Masson *et al.* 2002). In a regional sense, repeated landsliding will be a natural consequence if the sediment deposition processes that generate the preconditions for landslides persist over long periods of time. Cyclic conditions, such as glacial/interglacial transitions may also contribute to repeated landsliding on timescales similar to the cyclicity (Bryn *et al.* 2005). One of the key reasons for repeated landslides at a site specific scale is that the scars created by landslides often act as traps for subsequent sedimentation, leading to enhanced sedimentation rates and increasing the risk of further landslides. Thus contourites are preferentially deposited in landslide scars on the Norwegian margin, enhancing weak layer development within these scars (Bryn *et al.* 2005; Solheim *et al.* 2005). Similarly, landslide scars in the Canary islands are frequently the loci of subsequent volcanism, probably because the landslide removes some of the overburden and creates an easier path for magma to reach to the surface (Cantagrel *et al.* 1999). It is also possible that the debris left by a landslide, when loaded by subsequent volcanic products, can act as a weak layer for future landslides.

Landslide Processes

Submarine landslides can be sub-divided into a bewildering variety of types. However, in terms of volume of gravity-driven sediment transport in the ocean, only slides, debris flows and turbidity currents make a significant contribution. Debris avalanches are less significant in terms of total transport, but they pose a particular threat to human populations.

There is a generally accepted 'paradigm' that landslides in cohesive sediments evolve downslope from slide to debris flow to turbidity current through gradually increasing disintegration and entrainment of water (e.g. Mulder & Cochonat 1996; Ilstad *et al.* 2004; Bryn *et al.* 2005). However, this is probably an oversimplification in that some landslides travel many hundreds of km without appreciable transformation into turbidity currents, while others transform entirely into turbidity currents very close to source. In truth, the formation of large turbidity currents, in which a few hundred km^3 of (usually cohesive) sediment are rapidly mixed with much larger volumes of seawater, is a very poorly understood process (Talling *et al.* 2002).

Large landslides in continental margin sedimentary sequences are often complex events, and elements of slide, debris flow and turbidity current may all be evident in the aftermath of a single landslide. Often the slide scar will contain displaced but coherent slide blocks made up of sediments that have travelled only a short distance from source. Further downslope, the landslide deposit may show flow structures characteristic of debris flow processes. A turbidity current initiated by the landslide may travel hundreds of km beyond the obvious limit of the debris deposit, with no obvious connecting pathway or deposit. In this situation the correlation between turbidite and landslide can often only be established on the basis of precise dating, sedimentology and/or geochemical analysis (Pearce & Jarvis 1992; Wynn *et al.* 2002).

Theoretical and experimental studies have shown that landslides on slopes as low as 0.5° are only possible where excess pore pressures at the level of the detachment surface support a large fraction of the weight of the landslide mass, thus decreasing the effective stress and the friction with the underlying substrate (Iverson 1997; Kvalstad *et al.* 2005). High pore pressures can be created through rapid sediment deposition (especially in fine-grained sediments with low permeability), collapse of the sediment structure (so-called 'sensitive clays') due to earthquake shaking, or possibly due to melting of gas hydrates contained within the sediment.

Debris avalanches occur when a mass of cohesionless material, usually fragmented rock, moves downslope. In the subaerial realm, debris avalanches typically occur on slopes that range from 25° to near vertical; in this situation they can attain speeds as high as 100 m s^{-1} (360 km h^{-1}; Voight & Pariseau 1978). The most widely known debris avalanches in the submarine realm are those that occur on the flanks of volcanic islands (Moore *et al.* 1989; Watts & Masson 1995; Ollier *et al.* 1998; Masson *et al.* 2002), although they also occur in consolidated sedimentary rocks on active continental margins due to failure of steep slopes generated by tectonic processes (Hühnerbach *et al.* 2005). On the submarine flanks of volcanic islands such as the Canary islands, debris avalanche failure planes dip oceanward at 10° or less, suggesting that these avalanches are less energetic than their subaerial counterparts. The blocky character of submarine debris avalanche deposits probably reflects the friable nature of the volcanic material and the distance of transport (typically an order of magnitude greater than the biggest subaerial avalanches) rather than the speed of the emplacement process.

The Storegga Slide

As noted in a previous section, the Storegga slide, which occurred on the continental slope west of Norway around 8200 calendar years ago, is one of the largest and best-studied landslides on earth (Bugge *et al.* 1987, 1988; Haflidason *et al.* 2003; Haflidason *et al.* 2004; Bryn *et al.* 2005; Kvalstad *et al.* 2005).

The Storegga slide illustrates many key aspects of landslides on low angle continental slopes. These include:

1. Several (at least 4) distinct gently sloping (0.5–2°) failure planes parallel to the sedimentary bedding.
2. Steep (10–35°) headwall scarps separating the different glide plane levels.
3. Landslide debris showing clear evidence of brittle deformation preserved in the landslide scar.
4. Retrogressive behaviour.

An elegant geotechnical model for the Storegga slide was constructed by Kvalstad *et al.* (2005) who demonstrated that the best explanation for the slide required a combination of one or more weak layers, identified as marine clays (contourites) deposited during interglacial periods, and excess pore pressures developed as a result of the rapid sedimentation that loaded the Norwegian slope during intervening glacial periods. Evidence that high pore pressures existed in the area comes from measurements adjacent to the slide, where remnant high pressures can still be found. Modelling of the pore pressures found that excess pore pressure ratios of the order of 0.9 were required to cause failure. In simple terms this means that the strength of the sediment was reduced to about 10% of its normal level. There is clear evidence that instability only developed at certain levels in the sediment and that consequently most of the sediment pile was inherently stable. This is best seen on high-resolution seismic data that show faulted and fractured slide debris preserved within the slide scar (Kvalstad *et al.* 2005). It is also shown by the long term stability of the steep slide headwall scarps, which have stood with angles of up to 35° for over 8000 years since the excess pore pressure in the weak layers was released by the landslide.

The area of the Storegga slide shows a long-term history of sedimentation and landsliding that reflects glacial/interglacial cyclicity (Solheim *et al.* 2005). This leads to the conclusion that the state of stability that has characterized the Storegga slide area since the last landslide occurred is unlikely to change until the next interglacial/glacial cycle has been completed.

Canary Island Landslides

Large-scale landslides are a common feature of volcanic ocean islands such as Hawaii, the Canary islands and Reunion island (Moore *et al.* 1989, 1994; Watts & Masson 1995; Masson 1996; Ollier *et al.*

1998; Masson *et al.* 2002). The Nuuanu landslide, off Oahu in the Hawaiian islands, with an estimated volume of 5000 km^3, may be the largest single landslide on earth (Moore *et al.* 1989). Landslides on volcanic islands typically take two forms—debris avalanches and slumps (in the terminology used in this paper a slump is a type of slide). As defined by Moore *et al.* (1989), a debris avalanche is a relatively thin (0.4–2 km thick) landslide with a clear evacuated headwall and a distal train of blocky debris. Each debris avalanche appears to be a single event, at least in terms of geological time, and some show evidence for rapid and energetic emplacement. In contrast, a slump involves gradual, intermittent, downslope movement of a thick (up to 10 km) coherent block of the island flank.

The history of landslides in the Canary islands over the past one million years is now well understood. The bulk of landslide activity is associated with the youngest and most volcanically active islands of Tenerife, La Palma and El Hierro. On average, one landslide has occurred somewhere in the Canary islands every 100 000 years, although this figure masks an irregular distribution through time (Masson *et al.* 2002). The youngest landslide occurred on the island of El Hierro some 15 000 years ago. Most of the landslides are debris avalanches, with slumps only recognized on the youngest island, El Hierro, perhaps suggesting that this landslide style is a feature of early island development. A typical Canary island debris avalanche is marked by a near-vertical amphitheatre-shaped headwall on the island, an erosive chute on the upper part of the submarine island slope and a pile of avalanche debris at the foot of the steepest island slope, usually at 3000–4000 m waterdepth. This typical avalanche has a volume of 50–200 km^3, covers an area of a few thousand km^2 and has a run-out of 50–100 km. Glide planes at the base of the landslide are typically up to 10° on the upper slope, decreasing to less than 5° on the lower slope (Watts & Masson 1995; Gee *et al.* 2001; Watts & Masson 2001). Large accumulations of debris, such as seen north of Tenerife or west of La Palma, are clearly the cumulative result of several landslides rather than single larger events that have occurred in the past. Even some deposits thought to be the result of a 'single' landslide event (in geological time) show signs of more than one phase of emplacement (Watts & Masson 2001). It is notable that Canary island debris avalanches are an order of magnitude smaller than those on Hawaii, possibly reflecting the larger size and higher magma production rates of the Hawaiian islands or the more rapid development of instability on the relatively steeper Canary island slopes.

Canary island landslides identified as debris avalanches according to the definition of Moore *et al.* (1989) show a variety of structures that suggest elements of both debris avalanche and debris flow emplacement mechanisms. In addition, these landslides can also initiate turbidity currents that are capable of flowing considerable distances downslope. Sediment cores recovered from the deep Agadir Basin, about 300 km to the north of the islands, contain turbidite deposits which, based upon their mineralogy, geochemistry and age, are interpreted to be linked to Canary island landslides (Wynn *et al.* 2002; Wynn & Masson 2003). Specifically, a turbidite deposited at <“15 kyr is linked to the El Golfo landslide on El Hierro, while an older turbidite dated at *ca* 170 kyr is linked to the Icod landslide on Tenerife. Most turbidite deposits in the Agadir Basin are actually derived from the Moroccan continental margin to the east (Wynn *et al.* 2002), and show the typical smooth upward-fining grain-size profile that is typical of graded turbidite deposits. However, the two turbidites derived from Canary islands landslides show a stepped grain-size profile that appears to represent deposition from a series of 'mini-turbidites'. This pattern is interpreted to be the result of a multi-stage source landslide, as other potential causes, e.g. flow reflection, multiple pathways or pulses, can be ruled out (Wynn & Masson 2003). This hypothesis is also supported by the fact that similar turbidites linked to Hawaiian landslides show the same pattern of stacked mini-turbidites (Garcia 1996). Detailed sedimentological analysis of the Agadir Basin turbidites has revealed that their source landslides probably occurred in several retrogressive stages over a period of hours or days rather than weeks or months (Wynn & Masson 2003). Assessing the sedimentary record of deposits derived from these landslides is therefore critical when assessing their tsunamigenic potential, since it is clear that a series of smaller landslides spread over several hours will have a much smaller tsunami-building potential than a single large, instantaneous landslide.

Landslide-Generated Tsunamis

The generation and propagation of tsunamis resulting from earthquakes have been studied for the last 50 years and are now relatively well understood (Bardet *et al.* 2003). In contrast, the importance of tsunamis generated by landslides has only become widely recognized during the last fifteen years or so, when it became apparent that a landslide source could explain the unusual run-up distributions and propagation characteristics of certain particularly deadly tsunami, such as the 1998 PNG event (Ward 2001; Bardet *et *al.* 2003 and references therein; Okal & Synolakis 2004). However, the complexity and variability of submarine landslides means that we are still some way from a comprehensive understanding of the range of tsunamis that landslides are capable of producing. Modelling of landslide tsunamis has shown both that extreme wave heights of hundreds of metres might be possible (Ward & Day 2001; McMurty *et al.* 2004), but that models are sensitive to the geological input parametres and the hydrodynamic assumptions adopted in the model (Ward 2001; Haugen *et al.* 2005; Lovholt *et al.* 2005), with the result that poorly constrained model predictions will have large uncertainty.

Despite the variability of submarine landslides that might cause tsunamis, many such tsunamis show similar general characteristics. In particular, these tsunamis often have very large run-ups close to the landslide site but appear to propagate much less efficiently than earthquake tsunami, so have limited far-field effects (Okal & Synolakis 2004). This was exemplified by the 1998 PNG tsunami, where waves up to 15 m high affected a 20 km segment of coast, killing 2200 people (McSaveney *et al.* 2000) even though farther a field the tsunami was not a significant event (Okal & Synolakis 2004). This is a consequence of the relatively small source areas of most landslide tsunami (compared to the areas affected by large earthquakes) that leads to the generation of shorter wavelength waves. These are more prone to coastal amplification (increasing the local effect) and to radial damping (decreasing the distal effect). This contrasts with the lack of radial damping seen in earthquake tsunamis that are generated by elongate two-dimension sources; these tsunamis propagate perpendicular to the source fault with little radial spreading.

The tsunami generated some 8200 years ago by the Storegga slide off Norway is one of the best-understood landslide tsunamis. A well-preserved record of tsunami deposits on land in Norway, Scotland and the Faroe islands (Dawson *et al.* 1988; Bondevik *et al.* 1997, 2005) coupled with an unprecedented knowledge of the landslide processes (Bryn *et al.* 2003, 2005; Haflidason *et al.* 2003, 2004; Kvalstad *et al.* 2005), provides the basis for rigorous testing of tsunami models and allows the key landslide parametres to be identified (Bondevik *et al.* 2005; Haugen *et al.* 2005; Lovholt *et al.* 2005). The key findings are that landslide volume, velocity, initial acceleration, length and thickness all contribute to the determination of tsunami character. The best indicator of tsunamigenic potential is the product of volume and initial acceleration (Lovholt *et al.* 2005). An abrupt deceleration might also contribute to larger surface elevations. The slide length affects both the wavelength and the maximum surface elevation (Haugen *et al.* 2005), while the wavelength is also determined by the travel time or run-out distance of the slide. Submarine slides are normally clearly subcritical, i.e. the Froude number (the ratio of slide speed to the speed of wave propagation) is much less than one. This implies that the tsunami will run away from the wave-generating slide, limiting the build-up of the wave. Slides in shallow waters are more critical, since the speed of wave propagation is lower here. Moreover, shallower water normally means less distance to the coast and a shorter distance available for radial damping. In contrast, tsunamis generated by earthquakes are more critical when the seabed displacement occurs in deeper waters, as the initial wave (which in this case depends much less on the water depth) will become shorter and more dangerously amplified when propagating from deeper to shallower waters.

The Storegga slide is best modelled as a retrogressive slide, with a peak velocity of 25–30 m s^{-1} (Bondevik *et al.* 2005). The retrogressive slide, of total length L, is modelled as a train of N fixed block slides released at different times t, but moving with identical velocity distributions (Haugen *et al.* 2005). For simplicity, the blocks have the same thickness h and the same length L/N. Moreover, the time lag $\ddot{A}t$ between release of two adjacent blocks is assumed to be equal. For waves propagating in the same direction as the slide, increasing $\ddot{A}t$ increases the distance between the surface elevations caused by the individual block modules. This decreases the overlap and results in a smaller amplitude and longer wave. For small time lags, the wave remains smooth, but as $\ddot{A}t$ increases, the distances between the individual block modules become large and the discrete nature of the retrogressive slide starts to show. Eventually, when $\ddot{A}t$ is sufficiently large, the waves generated by the block modules are completely separated.

The waves moving in the opposite direction of the block modules also move in the same direction as the retrogressive process. Thus, small time lags will decrease the distance between the surfaces elevations caused by the individual block modules. Consequently, the overlap increases, resulting in a larger amplitude but shorter wave. When the time lag equals the time it takes for the wave to traverse a block module, i.e. $\ddot{A}t=L/N(gH)^{-1/2}$, where H is the water depth, the individual surface elevations interfere in a perfectly constructive manner and the amplitude is at its maximum. As the time lags are increased further, the waves moving in the opposite direction of the block modules exhibit the same dependence on $\ddot{A}t$ as described above for the wave moving in the same direction as the slide.

Models of the Storegga slide indicate that retrogression must have been rapid, taking less than one hour; slower retrogression cannot reproduce the observed tsunami run-up pattern. Bondevik *et al.* (2005) compared field observations with run-up heights deduced from new numerical simulations of the Storegga slide tsunami. The shape and volume of the slide were adjusted to new and more detailed reconstructions of the slide (Forsberg 2002; Haflidason *et al.* 2005). In the revised slide model the maximum thickness (400 m) of the slide is near the upper headwall and it becomes gradually thinner towards the slide front in the offshore direction. The slide is modelled as a box that is skewed and smoothed to reproduce the reconstructed morphology of the Storegga slide, and moves with maximum velocities of 20–35 m s^{-1}. All simulations use a run-out distance of 150 km. A linear, long wave numerical model simulates the tsunami. A maximum velocity between 25 and 30 m s^{-1} with a 25% reduction of the surface elevation due to landslide retrogression makes the best match to the field observations (Bondevik *et al.* 2005). According to a two-dimension retrogressive slide model (Haugen *et al.* 2005), such a reduction is obtained with a very little time lag, about 15–20 s between the sliding blocks. Velocities of 25–30 m s^{-1} are also supported by De Blasio *et al.* (2005), who find mean (centre of mass) slide velocities of 25–35 m s^{-1} in their run-out studies of the Storegga slide.

Ward & Day (2001) modelled a tsunami generated by a landslide on the west flank of La Palma island in the Canaries. These authors concluded that for a worst-case scenario an initial wave elevation of up to 900 m could be generated by a landslide of 500 km^3 volume that reached a peak velocity of 100 m s^{-1} after 2 min of travel, i.e. with an acceleration of almost 1 m s^{-2}. This tsunami would have the capability to cross the Atlantic, producing a 10–25 m high wave along the east coast of North America. However, Ward & Day (2001) use a linear dispersive model that does not describe nonlinear effects and wave breaking. Hence, the initial waves may be overestimated and the wave spectrum could contain too great a proportion of short wave components experiencing the strongest amplification and producing the highest run-up. We note that similar conclusions have been reached by other authors (e.g. Mader 2001) who provide estimates of wave elevation along the North American coast an order of magnitude smaller than those calculated by Ward & Day (2001).

As noted for the Storegga slide, tsunami wave elevation is strongly influenced by landslide volume (or by some combination of landslide frontal area, length and thickness), by peak velocity, and by initial acceleration. For the La Palma scenario Ward & Day (2001) used maximum possible values for these parametres, adopting a 'worst-case' geological scenario as their model input. However, geological understanding of Canary island landslides, especially the offshore deposits, provides a number of reasons for challenging the models of Ward & Day (2001). These include questions of landslide process, volume and velocity. In terms of landslide failure processes, the scars left on the islands give few clues. However, they do allow the slopes of the glide planes on which the slides occurred to be measured. Even the steepest of these, on the upper slope of the El Golfo landslide off the island of El Hierro, only reaches a maximum of 10°, decreasing to about 5° at the base of the island flank where the avalanche deposit begins (Gee *et al.* 2001). Further offshore, key features of the landslide deposits suggest that both debris avalanche and debris flow processes are involved in landslide emplacement. The question of what this debris avalanche morphology actually tells us about the emplacement process, and particularly about the velocity of the landslide, is clearly important. Unfortunately, the term debris avalanche immediately brings to mind the rapid, catastrophic rock avalanches that occur in mountain ranges on land (Voight & Pariseau 1978). These can reach velocities in the order of 100 m s^{-1} on slopes that range from 25° to near vertical (Plafker & Ericksen 1978; Voight & Pariseau 1978). However, it is difficult to envisage this type of process for Canary island debris avalanches, where the slide plane gradients are less than 10°. Here it seems likely that the debris avalanche morphology reflects the disintegration of a slide composed of brittle volcanic rock and therefore primarily reflects the non-cohesive character of the slide material. In this situation, it probably tells us little about the velocity of the landslide.

Nevertheless, general considerations of submarine landslide behaviour leads us to predict that Canary island landslides must be relatively energetic, so peak velocities in the order of 50 m s^{-1} need to be considered in tsunami modelling. This is supported by evidence from the Ritter island volcano collapse, where velocities of 40 m s^{-1} have been modelled on the basis of known landslide volume and tsunami run-up heights (Ward & Day 2003) and from offshore Hawaii, where debris avalanches that run-up the far side of the islands' flexural moat require velocities in the order of 80 m s^{-1}. Using the scaling arguments of Ward & Day (2003), Canary island landslides might be expected to have velocities intermediate between those of the above examples, although given that the glide plane for the Ritter island landslide (10–

25°) is considerably steeper than glide planes in the Canaries (5–10°), 50 m s^{-1} might be considered a maximum for Canaries examples. This is still well below the velocity at which the tsunami would propagate in the deep water west of the Canary islands, so little coupling between the landslide and the tsunami would occur. The initial acceleration of the landslide is even less well constrained than the peak velocity, but is just as critical in determining the tsunami-generating potential of a landslide (Haugen *et al.* 2005). Ward & Day (2001) used an extremely high acceleration, 0.83 m s^{-2}, in their La Palma model. This is 50 times higher than the acceleration that produced the best-fit model for the Storegga slide (Lovholt *et al.* 2005) and appears unlikely on a glide plane of 10° or less. To put it in context, a landslide acceleration of 0.83 m s^{-2} is about half of what would act on a subaerial block on a 10° plane with zero friction or resisting forces, such as the need to displace the water that forms the tsunami ($a=g \sin 10°=1.70$ m s^{-2}).

Questions regarding the volume of any future landslide from La Palma, and about the way in which it would fail, are perhaps the most critical in determining the size of the modelled tsunami. Ward & Day (2001) used a volume of 500 km^3 in their calculations with the assumption that this failed as a single block. However, this volume is two to three times bigger than a typical Canary island landslide, even though the area affected by their potential landslide is quite typical. This is because they assume a failure surface '2–3 km below the summit of the volcano' (Ward & Day 2001), despite the fact that reconstruction of other landslides on the Canary islands, supported by the offshore debris flow volumes, show their maximum thicknesses to be in the order of 1–1.5 km (Urgeles *et al.* 1997; Masson *et al.* 2002; Hurlimann *et al.* 2004). Most importantly, observations of turbidites associated with Canary island (and Hawaiian) landslides show that they did not fail as single blocks, and that the separation in time of the phases of the multi-stage failures was almost certainly sufficient to separate their tsunami-building potential.

In summary, we would suggest that many of the parametres used in the worst-case scenario for a future landslide from the island of La Palma, as modelled by Ward & Day (2001), have been pushed to, or often beyond, their feasible maximum values. Any future tsunami is thus likely to be smaller than the Ward and Day prediction. Furthermore, it needs to be recognized that the worst-case landslide is in any case an unlikely event, with a probability much smaller than the 1 in every 100 000 years re-occurrence time calculated overall for Canary island landslides. Nevertheless, we can be certain that future landslides will occur in the Canaries. It is also probable that such an event will generate locally devastating tsunamis, with run-ups exceeding anything seen in historical tsunamis elsewhere. However, the ability of such landslides to produce significant trans-oceanic tsunamis is much more questionable.

Prediction and Risk Assessment

The prediction of landslides in the submarine realm, where any precursor movements cannot generally be observed, is problematic—we know in general where landslides occur (and will occur) but we are far from being able to provide reliable forecasts of individual events, especially where the final trigger is likely to be a transient event, such as an earthquake, which in itself cannot be predicted. On land, areas of particular risk, e.g. the west flank of La Palma in the Canary islands, can be monitored using seismometres or GPS arrays to warn of ground movement, but it is not practical to monitor all but the most high risk areas onshore, and almost impossible to monitor offshore areas. On human timescales, understanding how seabed exploitation, such as oil and gas exploration and production, might contribute to slope failure and landsliding is of current importance. The huge effort made to understand the Storegga slide, and the cost of this effort, demonstrate that even past landslides can be a significant financial hazard in the offshore industry.

The 2004 Indian Ocean tsunami was a reminder of the power of this natural phenomenon and that such events, created by landslides as well as earthquakes, have a worldwide distribution. Early warning systems for tsunamis are useful on a short time scale. On a longer time scale, we can only attempt to manage the consequences. Here, a probabilistic analysis of tsunami hazard is essential for estimating the potential losses and risk to human life and infrastructure along the coastline. The development of a Probabilistic Tsunami Hazard Analysis (PTHA) tool should follow the same lines as the Probabilistic Seismic Hazard Analysis (PSHA), which has become standard practice in the evaluation and mitigation of seismic hazards to structures, infrastructure and lifelines. The ability of PSHA to condense the complexities and variability of seismic activity into a manageable set of parametres greatly facilitates the design of effective seismic resistant buildings and infrastructure. The PTHA tool should be designed to provide a synopsis of the tsunami hazard along entire coastlines and identify the specific tsunami source regions that most strongly influence the tsunami hazard at each site on the coastline.

Because submarine landslides are rare on a human timescale, unpredictable, and almost impossible to observe and instrument, numerical modelling is seen as one of the key ways forward, both for understanding the landslides themselves and for the prediction of landslide tsunamis. Aspects of modelling are discussed in the following two sections.

Numerical Modelling of Landslide Dynamics

Submarine landslides may have huge dimensions and long run-out distances. Modelling the entire three-dimension problem is a huge computational task, and usually has to be reduced to a two-dimension problem through depth averaging or through restriction to cross sections to save computation resources. The loss of information about the vertical profiles of the velocity and density is usually insignificant, in particular for large-scale events. However, even a two-dimension simulation is a non-trivial task for large landslides. If it is known that the lateral spreading of the flowing mass is weak or limited, the flow evolution in the transversal direction may be neglected and one-dimension models may be applied.

Models should preferably describe the multi-layer structure of a submarine landslide with a dense debris flow at the bottom and a dilute turbidity current (suspension flow) above it. Hence, the vertical density variations and the associated variation of the mechanical properties should be taken into account. The dense debris flow is often considered either as a saturated mass of cohesionless material, or as a visco-plastic material where no deformation takes place until a specified stress is applied to the material, after which deformation is driven by the excess of the stress beyond the yield stress. An example is the frequently used Bingham fluid model, describing a viscous Newtonian fluid combined with a yield stress. Such a fluid moves as a dense plug flow riding on top of a shear flow. Material properties, including clay rheology, are of great importance to the flow dynamics and travel distance for the majority of events. The material properties also affect the rate of erosion and channelization. In the dense layer, energy dissipation is due to particle–particle interaction in granular flows with high particle concentrations, and to viscosity for more fluidized flows with lower particle concentration. Clay slurries exhibit yield strength threshold behaviour and viscosity, not only for (visco-plastic) shear flows, but also in extensional flows.

In modelling of failure and rupture, the separation of two material volumes that were originally adjacent poses specific problems. In simulating the evolution of a sliding block at the escarpment during the retrogressive phase of the Storegga slide, the progressive decay of the original slab into wedge-shaped blocks had to be reproduced. Two key factors in the model were responsible for the emergence of shear bands at the correct places (i.e. as observed in high-resolution profiles across the landslide), namely the assumption of shear softening and the high-resolution grid. All time- and location-dependent material properties such as the yield strength had to be advected with the flow, requiring substantial computing resources.

For engineering purposes, knowledge of impact loads is needed for both granular, visco-plastic and suspension flows. Depth-averaged (hydraulic) models may simulate high-Froude number granular flows against obstacles quite well (applying appropriate shock-capturing numerical schemes). For suspension flows with Froude numbers close to 1, the assumption of hydrostatic pressure distribution in the hydraulic models is poor in the head of the current where the largest pressures have to be expected, and complex two-dimension or three-dimension models may have to be used. The situation is less clear for visco-plastic flows because the rheology is only partially known, and because determination of yield surfaces will be involved. Today, it is not known whether two-dimension (depth-averaged) models are applicable when the yield strength is important.

Numerical Modelling of Landslide Tsunamis

The quantitative description of gravity mass flows poses a series of nontrivial problems linked to stability, rheology, disintegration and mixing with water during landslide propagation. In addition, modelling of the tsunami, in particular close to the source region, requires an approach different to that used in modelling tsunamis generated by earthquakes. The open sea propagation of earthquake tsunamis is most commonly described by the shallow water equations that are the simplest member of the class of depth integrated long wave equations. Even though errors due to frequency dispersion may accumulate in trans-oceanic propagation and that certain phenomena in shallow water as well as details in the generation are lost, the shallow water equations are generally adequate for this purpose. Moreover, the final seabed elevations from the earthquake are transferred to initial elevations on the ocean surface. As a rule this simple practice does suffice.

For submarine slides the corresponding simple source model is an equivalent sink/source distribution with prescribed shape and motion. However, this is indeed a crude model for submarine slides, for which the interaction between the solid constituents of the flow and the fluid may be crucial both for the slide dynamics and the generation of the tsunami. For rock-type slides depth integrated wave equations may still be used in combination with corresponding slide models, maybe with parametreization of features like exchange of fluid between the slide body and the surrounding fluid. On the other hand, mud type mass gravity flows will entrain water, and produce turbulence and large vortices, while form drag may have a crucial influence on the shape and dynamics of the mudflow. For slides with a subaerial origin the impact in water and progression in shallow depths involve wave breaking and huge wave resistance (momentum exchange with surface gravity waves in the water). Another challenge is the retrogressive slide dynamics, where the tsunami generation by a series of smaller slide masses clearly calls for hydrodynamic analysis and modelling beyond long wave assumptions. In short, a proper quantitative description of mass gravity flows and their tsunami generation requires general hydrodynamic computational tools, like Navier–Stokes type models, but with modified rheology for the denser types of slide masses. Today, three-dimension models of this kind are so computationally demanding that simulations with sufficient resolution are difficult to perform even on the most powerful computers.

Assuming that an appropriate model is available for the mass gravity flow and the generation of surface waves, the far-field tsunami propagation still has to be handled with some set of depth integrated, long wave equations, that must be coupled with the more complex model for the source region. In many cases the vicinity of the slide will be dominated by flow structures associated with vortices and turbulence that cannot be conveyed properly to a depth integrated model. In comparison to the energy and momentum following the tsunami, these features are advected much more slowly by the material fluid velocity. Hence, the simplest option would be to employ a large source region that allows the transition of pure waves to the depth integrated far field. However, this will be difficult to achieve due to computational limitations. Presumably, the wave motion must be identified and extracted in the vicinity of the slide, which is far from straightforward.

Mass gravity flows frequently occur in the aftermath of an earthquake, causing tsunamis additional to and much more localized than those from the earthquake itself, but at the same time potentially very destructive (Bardet *et al.* 2003; Okal & Synolakis 2004). Moreover, the wavelengths may occasionally be much shorter than for tsunamis of seismic origin, which implies that a dispersive model must be used (Lynett *et al.* 2003). The complete modelling of the mass gravity flow and the generation and propagation of the tsunami may require an optimal application and combination of a diversity of hydrodynamic models. Both the combination of computational tools and the need to parallelize the heavy computations point to a domain decomposition strategy. Such techniques, together with related techniques on nesting have been applied for a while, but more development and testing are needed, in particular for the combination of genuinely different models.

Landslides are ubiquitous features of submarine slopes in all geological settings and at all water depths. Because of their unique geological settings, submarine landslides can be up to two orders of magnitude bigger than those that occur on land. Hazards related to such landslides, ranging from destruction of offshore facilities to collapse of coastal facilities and the generation of tsunamis, therefore occur widely.

Most submarine slopes are inherently stable and require external, often transient effects to trigger landslides. Elevated pore pressures, related to processes such as earthquake shaking or rapid sedimentation appear to be the critical factor in most submarine landslides. Such pore pressures support part or all of the weight of the overlying sediment, thus lowering the frictional resistance to landsliding. When concentrated in specific geological layers, they create weak layers, which fail in a characteristic bedding plane parallel style. The Storegga slide, off western Norway, is a classic example of this type of landslide showing several distinct gently sloping failure planes parallel to the sedimentary bedding, steep headwall scarps separating the different glide plane levels, retrogressive landslide behaviour, and clear evidence that most of the landslide mass deformed by brittle deformation.

Landslides on the flanks of oceanic volcanic islands, such as the Canary islands, have received much attention because of their potential tsunamigenic behaviour. While it is certain that such landslides have occurred in the past and will again occur in the future, the causes and triggers of flank failure are relatively poorly understood. The best evidence suggests that these landslides are retrogressive and occur over a period of hours or days, rather than the instantaneous failure of a single coherent block. It therefore seems unlikely that simple models that treat these landslides as single rapidly moving blocks can accurately predict their tsunamigenic potential. In any case, it needs to be remembered that any worst-case scenario model, such as that produced by Ward & Day (2001) for La Palma in the Canary islands, which maximizes every model parametre, has a low probability of occurrence in the real world.

The precise forecasting of submarine landslides is still largely beyond our ability—areas at risk from landslide are relatively easy to identify, but we are far from being able to forecast the occurrence of individual events. Numerical modelling of landslides and related hazards such as tsunamis is one of the most obvious ways in which our understanding can be improved. Such models can be used to help estimate the potential losses and risk to human life and infrastructure due to landslides that in turn can be used to construct mitigation strategies.

Challenges of Devastating Indian Landslides

It is all in the books that landslides are among the major hydro-geological hazards that affect large parts of India, especially the Himalayas, the northeastern hill ranges, the Western Ghats, the Nilgiris, the Eastern Ghats and the Vindhyas, in that order. In the Himalayas one could find landslides of every name, fame and description. India's northeastern region, the Darjeeling district of West Bengal, Sikkim, Tripura, Meghalaya, Assam, Nagaland and Arunachal Pradesh are all landslide-prone. In Uttarakhand also, it would therefore be inappropriate to see the Kedarnath tragedy of 16–17 June 2013, as merely an isolated event frozen in time and space.

Whenever landslide disasters strike, we rush to lean on fixed ideas in our minds. From the school days we are tutored that events like landslides and earthquakes are only to be regarded as nature's safety valves because we live on the surface of an unfinished planet. The fragile ecology, immature geology, meandering rivers, snow bodies, climatic variations and cloudbursts of the Himalayas are after all our inheritance without choice. For centuries, landslides have come and gone, and these can be explained by recounting a long list of causative factors. If and when our justification is not good enough, there is climate change to buttress our argumentation. But, by ignoring human violence against nature, we only speak the half truth. Let us always remember that 'a scientific truth does not triumph by convincing its opponents and making them see the light, but rather because its opponents eventually die and a new generation grows up that is familiar with it' [Max Planck].

Those who make off-the-cuff conclusions must know that the science of landslides has no contract with their way of thinking. It demands honest and comprehensive scientific studies. We can understand landslides only by elucidation of landslide boundary-shears, concurrent monitoring of time-dependent piezometric pressures, surface and subslope displacements and mapping of ground deformations and shear zones, plus behavioural studies of associated human settlements form an integral part of the landslide investigation. Sadly, success will continue to elude us so long as scientific landslide investigation does not precede landslide remediation. We have not been able to fix the landslides not because of lack of expertise or technology, but because we never had the will or direction to do so.

It has almost become ritualistic to name cloudburst to explain away cataclysmic floods and devastating landslide events, without even attempting to understand the slope dynamics in the ecological theatre of nature. We did so to explain the great Alaknanda tragedy of 20–21 July 1970 in Uttarakhand on the premise that the previous maxima of 200 mm rainfall recorded at Joshimath on 28 September 1924 was crossed by a new high of 212.8 mm. Further probe removed a layer to show that the tragedy was caused by the bursting of a landslide dam. The formation of the landslide dam on Alaknanda was then traced to the enormous sediment load brought by Patalganga. And this huge sediment load was in turn traced to numerous landslides in the Patalganga valley. Further, it became evident that these landslides themselves were the result of neglect, misuse and abuse of our lands for decades on end. But for the ecological neglect, the Alaknanda floods would not have hit the headlines.

We are yet again stumped by the ghastly Kedarnath tragedy and cajoled by the very same reason – cloudburst, which is in fact no more than the most visible trigger at the tipping point. Could we have anticipated the trouble? The answer is no, because we had neither fail-safe instrumentation nor real-time vigil on our glaciers, glacial lakes, moraine accumulations, dormant and active landslides, rivers and their tributaries and unsafe housing stock. We plead for zero tolerance against mindless urbanization, but suffer it instead. We have mapped landslide hazards on the pilgrim routes many times over, but never placed a single user-friendly validated map in the hands of disaster managers. We should do it now.

We should not continue to ignore the gross disconnect between our scientific discourse and our approach to hazard-mapping. In scientific discourse, we dread factors such as climate change, exceptional rain, receding of glaciers, bursting of glacial lakes, poor road alignments, non-engineered constructions, earthquake-induced landslides, and overtopping of dams. However, in the case of hazard-mapping, we disregard all these factors and only account for lithology, structure, slope morphometry, relative relief, land use/land cover and hydro-geological condition India needs large-scale, validated and user-friendly hazard maps based on a scientific understanding of the multitude of factors, both natural and human induced.

The main reason why the natural landslide hazards are turning into man-made disasters is because people have not only moved in large numbers to the remotest of the mountain slopes where no one ever lived before, but the violence they have unleashed against nature is unprecedented. There is a Chinese saying that 'a man who removes a mountain begins by carrying away small stones'. We have been removing, not small stones, but mountains of rocks for building townships, roads, dams and reservoirs. Many of the landslides we know are the result of these very actions, and they in turn remove from slopes incredibly large amount of sediments, loading rivers, silting reservoirs and creating new land masses in the sea. Little do we realize that when a slope gets robbed of one inch of its soil cover, Nature may take nearly 1000 years to replenish it! It is time therefore to revive the Chipko Movement and reverse the trend of slope degradation.

One vexing question which often haunts us is whether a landslide can be predicted and a landslide disaster averted? In Uttarakhand, some of the landslides occur annually. We do not need any rocket science to predict them; simple slope instrumentation and monitoring would do. Similarly, mountain slopes supporting human habitat with visible signs of instability like tilting of trees, bulging of retaining walls and widespread ground subsidence are already on the verge of failure. What more early warning do we need to predict a landslide in such situations? It is a scientifically proven fact that even the first time landslides are predictable provided we probe deep enough to arrive at the bottom of the truth, through studies, instrumentation and monitoring. Like human beings, a slope also has a heart that beats! Let us recall Terzaghi, who more than six decades ago said that 'If a landslide comes as a surprise to eye witnesses, it would be more accurate to say that the observers failed to detect the phenomena which preceded the slide.'

Today, we have the knowledge, tools and experience we need to predict and avert most, if not all, landslides. By tapping the phenomenal power of geotechnology, instrumentation, remote sensing, integrated GPS and information communication systems, we can monitor unstable areas in real time even during unfavourable weather conditions. It is time therefore to launch selected mission-mode projects to initially cover timely prediction of (a) possible reactivation of major old, dormant and seasonal landslides, (b) landslides and floods due to bursting of glacial lakes, (c) flash floods due to bursting of landslide dams, (d) first-time landslides in urban and strategically important areas falling in the zone of exceptional landslide hazard and (e) rockfalls. But, why is this not happening?

The criteria for early warning against landslides we use must be credible. The direct connection between 'incidence' of a landslide and 'rainfall' may look both obvious and simple, and may even work in cases where ground conditions are already bad enough and rainfall exceptional. There is a strong case to position monitoring stations to advance on-line rainfall forecasting procedures using digital radar data and an on-line run-off forecasting procedure based on space techniques to enhance lead time. The early warning criteria we aim should be rooted in holistic and concurrent interpretation of real-time rainfall records, seismic records, spatial piezometric variations, slope surface and subsurface movements and movement rates on discrete boundary shears, runout effects and other collateral threats in the catchment and on the higher slopes. We should refrain from over-simplifying the criteria for early warning to minimize bogus forecasts and it must be continuously put to test. We must prepare ourselves to effectively utilize every second of the available lead time.

We have all agreed time and again that landslide disaster management should be integrated with development planning. The vast potential for hydro-power in Uttarakhand is in a sense a big blessing, but the way it is being exploited is a curse as hydro-power schemes are no longer environment-friendly and power generation is no longer based on natural flows and sound engineering. Unless safety issues appear continuously on RADAR, mega projects like the Tehri dam will always keep us on tenterhooks.

Dozens of landslides in India, like the one at Kaliasaur on the Srinagar–Rudraprayag highway, are quite old. We should fix them once for all. A township on a landslide infested mountain slope can be best tackled by looking at the stability of the mountain as a whole rather than frittering away the resources in fixing landslides affecting individual buildings. We need breed the culture of truth-seeking rather than data-seeking nature of landslide investigations. Our reports and papers by hindsight reflect more of perceptions inspired by loyalty to the accepted trends than science. In many cases, truth eludes us because vital field evidences get erased even before landslide investigations begin.

Landslide Tsunamis

The Other Kind of Tsunami

By now everyone on Earth knows about tsunamis—the dreadful tsunamis of 2004 and 2011 have seen to that, especially for people unfamiliar with the earlier tsunamis of 1946, 1960 and 1964. Those tsunamis were of the ordinary type, seismic tsunamis caused by earthquakes that suddenly lift or drop the seafloor. But a second type of tsunami can arise from landsliding with or without an earthquake, and shorelines of every kind, even lakes on land, are susceptible. Landslide tsunamis are harder to predict, harder for scientists to model and harder to defend against.

Landslide Tsunamis and Earthquakes

Landslides of various kinds can push water around. Mountains may crumble to the sea, as the song goes. Mudslides may plop into lakes and reservoirs. And land that lies entirely beneath the waves may fail. In all cases, the landslide material displaces water, and the water responds in very large waves that spread rapidly out in all directions.

Many landslides occur during earthquakes, so landslides can complicate seismic tsunamis. The Grand Banks earthquake in eastern Canada on 18 November 1929 was tolerable, but the ensuing tsunami killed 28 people and ruined the economy of southern Newfoundland. The landslide was quickly detected by the fact that it broke 12 submarine cables linking Europe and America with communications traffic.

The role of landslides in tsunamis has become more important as tsunami modeling has advanced. The deadly Aitape tsunami in Papua New Guinea on 17 July 1998 was preceded by an earthquake of magnitude 7, but seismologists couldn't make the seismic data match the tsunami observations until seafloor surveys later showed that a large submarine landslide was also involved. Now awareness has been raised.

Today the best advice is to beware of a tsunami *any* time you experience an earthquake near *any* body of water. Alaska's dire Lituya Bay, a steep-walled fjord on a major fault zone, has been the site of several stupendous landslide tsunamis related to earthquakes including the largest one on record. Lake Tahoe, high in the Sierra Nevada between California and Nevada, is prone to both seismic and landslide tsunamis.

Human-Caused Tsunamis

In 1963, a massive landslide pushed some 30 million cubic metres of water over the new Vajont Dam, in the Italian Alps, which killed some 2500 people. The filling of the reservoir destabilized the adjoining mountainside until it gave way. Amazingly, the reservoir designers were attempting to let the mountainside collapse gently by manipulating the water level. Dave Petley, writer of the Landslide Blog, doesn't use the word tsunami in his description of this manmade tragedy, but that's what it was.

Prehistoric Megatsunamis

Recently with the improved maps of the world's seafloor, we have found evidence suggesting truly gigantic disturbances that must have created landslide tsunamis equal to today's worst events. Like the supposed threat of "supervolcanoes" based on the great size of ancient volcanic deposits, the idea of impending "megatsunamis" has gotten a lot of credulous attention.

Very large seafloor landslides could occur in many places, where they could have produced tsunamis. Consider the fact that rivers are constantly depositing sediment on the continental shelves on the edge of every continent. At some point there will be one sand grain too many, and a runaway landslide over the edge of the shelf could move a lot of material beneath a lot of water. If a distant earthquake isn't the trigger, a large local storm could be.

Also to be considered is the long-term climate, including ice ages. Rising water temperatures or falling sea levels that accompany different stages of an ice age could destabilize the delicate methane hydrate deposits in subarctic regions. That kind of slow destabilization is one common explanation for the enormous Storegga Slide in the North Sea off Norway, which left widespread tsunami deposits on surrounding lands about 8200 years ago. Given that sea level has been steady ever since, we can discount the possibility that a repeat slide is imminent even though the average ocean temperature is likely to rise with global warming.

Another postulated tsunami mechanism is the collapse of volcanic islands, which are generally considered to be more fragile than continental rocks. There are large chunks of Molokai and other Hawaiian islands found far away on the Pacific Ocean floor, for instance. Similarly, the volcanic Canary and Cape Verde islands in the North Atlantic are known to have collapsed at times in the past.

Scientists who modeled these collapses got a lot of press a few years ago when they suggested that eruptions on these islands could cause them to fall apart and raise truly killer waves all around the Pacific or Atlantic shoreline. But there are compelling arguments that nothing like this is likely today. Like the thrilling threat of "supervolcanoes," megatsunamis would be foreseeable many years in advance.

Strong Ground Motion

Seismologists usually define strong ground motion as the strong earthquake shaking that occurs close to (less than about 50 km from) a causative fault. The strength of the shaking involved in strong ground motion usually overwhelms a seismometre, forcing the use of accelerographs (or strong ground motion accelerometres) for recording.

As seismic instruments (and accelerometres in particular) become more common, it becomes necessary to correlate expected damage with instrument-readings. The old Modified Mercalli intensity scale (MM), a relic of the pre-instrument days, remains useful in the sense that each intensity-level provides an observable difference in seismic damage.

After many years of trying every possible manipulation of accelerometre-time histories, it turns out that the extremely simple peak ground velocity (PGV) provides the best correlation with damage. PGV merely expresses the peak of the first integration of the acceleration record. Accepted formulae now link PGV with MM Intensity. Note that the effect of soft soils gets built into the process, since one can expect that these foundation conditions will amplify the PGV significantly.

ShakeMap systems tie all of this together into a useful product. Some systems use seismometres and accelerometres to produce a near-instantaneous map of expected MM Intensities after a significant earthquake. As well, people can send in their observations of earthquake effects to help fill in the maps, which can help disaster-relief teams and other agencies.

The science of strong ground motion also deals with the variations of fault rupture, both in total displacement, energy released, rupture velocity, etc.

5

Peak Ground Acceleration

Peak ground acceleration (PGA) is a measure of earthquake acceleration on the ground and an important input parametre for earthquake engineering, also known as the design basis earthquake ground motion (DBEGM). Unlike the Richter and moment magnitude scales, it is not a measure of the total energy (magnitude, or size) of an earthquake, but rather of how hard the earth shakes in a given geographic area (the *intensity*). The Mercalli intensity scale uses personal reports and observations to measure earthquake intensity but PGA is measured by instruments, such as accelerographs, and it generally correlates well with the Mercalli scale. The peak horizontal acceleration (PHA) is the most commonly used type of ground acceleration in engineering applications, and is used to set building codes and design hazard risks. In an earthquake, damage to buildings and infrastructure is related more closely to ground motion, rather than the magnitude of the earthquake. For moderate earthquakes, PGA is the best determinate of damage; in severe earthquakes, damage is more often correlated with peak ground velocity.

Geophysics

Earthquake energy is dispersed in waves from the epicentre, causing ground movement horizontally (in two directions) and vertically. PGA records the acceleration (rate of change of speed) of these movements, while peak ground velocity is the greatest speed (rate of movement) reached by the ground, and peak displacement is the distance moved. These values vary in different earthquakes, and in differing sites within one earthquake event, depending on a number of factors. These include the length of the fault, magnitude, the depth of the quake, the distance from the epicentre, the duration (length of the shake cycle), and the geology of the ground (subsurface). Shallow-focused earthquakes generate stronger shaking (acceleration) than intermediate and deep quakes, since the energy is released closer to the surface.

Peak ground acceleration can be expressed in *g* (the acceleration due to Earth's gravity, equivalent to g-force) as either a decimal or percentage; in m/s^2 ($1\ g = 9.81\ m/s^2$); or in Gal, where 1 Gal is equal to $0.01\ m/s^2$ ($1\ g = 981$ Gal).

The ground type can significantly influence ground acceleration, so PGA values can display extreme variability over distances of a few kilometres, particularly with moderate to large earthquakes. The varying PGA results from an earthquake can be displayed on a shake map. Due to the complex conditions affecting PGA, earthquakes of similar magnitude can offer disparate results, with many moderate magnitude earthquakes generating significantly larger PGA values than larger magnitude quakes.

During an earthquake, ground acceleration is measured in three directions: vertically (V or UD, for up-down) and two perpendicular horizontal directions (H1 and H2), often north-south (NS) and east-west (EW). The peak acceleration in each of these directions is recorded, with the highest individual value often reported. Alternatively, a combined value for a given station can be noted. The peak horizontal ground acceleration (PHA or PHGA) can reached by selecting the higher individual recording, taking the mean of the two values, or calculating a vector sum of the two components. A three-component value can also be reached, by taking the vertical component into consideration also.

In seismic engineering, the effective peak acceleration (EPA, the maximum ground acceleration to which a building responds) is often used, which tends to be $\frac{2}{3}$ – $\frac{3}{4}$ the PGA.

Seismic Risk and Engineering

Study of geographic areas combined with an assessment of historical earthquakes allows geologists to determine seismic risk and to create seismic hazard maps, which show the likely PGA values to be experienced in a region during an earthquake, with a probability of exceedance (PE). Seismic engineers and government planning departments use these values to determine the appropriate earthquake loading for buildings in each zone, with key identified structures (such as hospitals, bridges, power plants) needing to survive the maximum considered event (MCE).

Damage to buildings is related to both peak ground velocity and PGA, and the duration of the earthquake – the longer high-level shaking persists, the greater the likelihood of damage.

Comparison of Instrumental and Felt Intensity

Peak ground acceleration provides a measurement of *instrumental intensity*, that is, ground shaking recorded by seismic instruments. Other intensity scales measure *felt intensity*, based on eyewitness reports, felt shaking, and observed damage. There is correlation between these scales, but not always absolute agreement since experiences and damage can be affected by many other factors, including the quality of earthquake engineering.

Generally speaking,

- 0.001 g (0.01 m/s^2) – perceptible by people
- 0.02 g (0.2 m/s^2) – people lose their balance
- 0.50 g – very high; well-designed buildings can survive if the duration is short.

Correlation with the Mercalli Scale

The United States Geological Survey developed an Instrumental Intensity scale which maps peak ground acceleration and peak ground velocity on an intensity scale similar to the felt Mercalli scale. These values are used to create shake maps by seismologists around the world.

Instrumental Intensity	*Acceleration (g)*	*Velocity (cm/s)*	*Perceived Shaking*	*Potential Damage*
I	< 0.0017	< 0.1	Not felt	None
II-III	0.0017 - 0.014	0.1 - 1.1	Weak	None
IV	0.014 - 0.039	1.1 - 3.4	Light	None
V	0.039 - 0.092	3.4 - 8.1	Moderate	Very light
VI	0.092 - 0.18	8.1 - 16	Strong	Light
VII	0.18 - 0.34	16 - 31	Very strong	Moderate
VIII	0.34 - 0.65	31 - 60	Severe	Moderate to heavy
IX	0.65 - 1.24	60 - 116	Violent	Heavy
X+	> 1.24	> 116	Extreme	Very heavy

Other Intensity Scales

In the 7-class Japan Meteorological Agency seismic intensity scale, the highest intensity, Shindo 7, covers accelerations greater than 4 m/s^2 (0.41 *g*).

PGA Hazard Risks Worldwide

In India, areas with expected PGA values higher than 0.36*g* are classed as "Zone 5", or "Very High Damage Risk Zone".

Notable Earthquakes

PGA single direction (max recorded)	PGA vector sum (H1, H2, V) (max recorded)	Mag	Depth	Fatalities	Earthquake
2.7g	2.99 g	9.0	30 km	>15000	2011 Tōhoku earthquake and tsunami
2.2g		6.3	5 km	185	February 2011 Christchurch earthquake
2.13g		6.4	6 km	1	June 2011 Christchurch earthquake
	4.36g	6.9/7.2	8 km	12	2008 Iwate-Miyagi Nairiku earthquake
1.7g		6.7	19 km	57	1994 Los Angeles earthquake
	1.47g	7.1	42km	4	April 2011 Miyagi earthquake
1.26g		7.1	10 km	0	2010 Canterbury earthquake
1.01g		6.6	10 km	11	2007 Chūetsu offshore earthquake
1.01g		7.3	8 km	2,415	1999 Jiji earthquake
1.0g		6.0	8 km	0	December 2011 Christchurch earthquake
0.8g		6.8	16 km	6,434	1995 Kobe earthquake
0.78g		8.8	23 km	521	2010 Chile earthquake
0.6g		6.0	10 km	143	1999 Athens earthquake
0.51g		6.4		612	2005 Zarand earthquake
0.5g		7.0	13 km	92,000-316,000	2010 Haiti earthquake
0.438g		7.7	44 km	27	1978 Miyagi earthquake (Sendai)
0.367g		5.2	1 km	9	2011 Lorca earthquake
0.25 - 0.3g		9.5	33 km	1,655	1960 Valdivia earthquake
0.24g		6.4		628	2004 Morocco earthquake
0.18g		9.2	23 km	143	1964 Alaska earthquake

Debris Flow

Debris flows are fast moving, liquefied landslides of mixed and unconsolidated water and debris that look like flowing concrete. They are defined by their non-newtonian flow dynamics, and behave as Bingham plastics. This characteristic can lead to the formation of levees at the margins of unconstrained debris flows as the margins of the flow freeze. They are differentiated from mudflows by their coarser and more poorly sorted sediment load. Flows can carry material ranging in size from clay to boulders, and may contain a large amount of woody debris such as logs and tree stumps.

Figure: *Debris flow channel with deposits left after **2010** storms in Ladakh, NW Indian Himalaya. Note coarse bouldery levees on both sides of the channel, and poorly sorted rocks on the channel floor.*

Figure: *Scars formed by debris flow in greater Los Angeles during the winter of **1968-1969**. The photograph was taken within several months of the debris flows occurring.*

Flows can be triggered by intense rainfall, glacial melt, or a combination of the two. Speed of debris flows can vary from 5 km/h to up to 80 km/h in extreme cases. Volumes of material delivered by single events vary from less than 100 to more than 100,000 cubic metres. Variables considered important in debris flow initiation include slope angle, available loose sediment, and degree of land disturbance by activities such as forest harvesting. Debris flows are often more frequent following forest and brush fires, as experience in southern California clearly demonstrates. Debris flows are extremely destructive to life and property, and claim thousands of lives world-wide in any given year. They are a particular problem in steep mountainous areas subjected to intense rainstorms, and have received particular attention from researchers in Japan, Western USA, Western Canada, New Zealand, the European Alps, and Kazakhstan.

True debris flows are very sediment dense, typically > 60 % sediment by volume. Less concentrated flows would be hyperconcentrated or fluvial.

Features and Behaviour

Figure: *Ancient debris flow deposit, Resting Springs Pass, California.*

Debris flows generally form when unconsolidated material becomes saturated and unstable, either on a hillslope or in a stream channel. Flows are accelerated downhill by gravity and tend to follow steep mountain channels. The front, or 'head' of debris flow often contains a great deal of coarser material such as boulders and logs. Trailing behind this frontal lobe is the less viscous, main part of the flow that contains sand, silt and clay. Debris flows eventually become 'thinner', or less viscous, muddy flood waters as they deposit their coarser components in areas of reduced gradient. Debris flows tend to move in pulses, or discrete surges, as friction or other barriers are overcome during the flow. Sometimes earlier pulses or previous debris flow deposits form levees which confine the flow until they are breached by later, larger flows. The presence of older levees (a.k.a. lateral deposits) provides some idea of the magnitudes of previous debris flows in a particular area, and through dating of trees growing on such deposits, may indicate the approximate frequency of destructive debris flows. This is important information for land development in areas where debris flows deposit material, known as a debris cone or colluvial fan. The big debris flow or landslide is called *yamatsunami* (q\%mâl), literally *mountain tsunami* in Japan.

Types

Other geological flows can also be described as debris flows, though are typically given more specific names. These include:

Lahar

A lahar is a debris flow related in some way to volcanic activity, either directly as a result of an eruption, or indirectly by the collapse of loose material from the flanks of a volcano. A variety of factors may trigger a lahar, including melting of glacial ice due to volcanic activity or climate change, intense rainfall on loose pyroclastic material, or the out bursting of a lake that was previously dammed by pyroclastic or glacial material. The word lahar is of Indonesian origin, but is now routinely used by geologists world-wide to describe volcanogenic debris flows.

Jökulhlaup

A jökulhlaup is a debris flow that originates from a glacial outburst flood. Jökulhlaup is an Icelandic word which refers specifically to floods having a glacial trigger. In the case of Iceland, many such floods are triggered by sub-glacial volcanic eruptions, since Iceland sits atop the Mid-Atlantic Ridge. Elsewhere, a more common cause of jökulhlaups is the breaching of ice-dammed or moraine-dammed lakes. Such breaching events are often caused by the sudden calving of glacier ice into a lake, which then causes a displacement wave to breach a moraine or ice dam. Downvalley of the breach point, a jökulhlaup may increase greatly in size through entrainment of loose sediment and water from the valley through which it travels. Travel distances may exceed 100 km.

Theories and Models of Debris Flows

Debris flow is an example of granular convection which occurs in many different environments and scales. Numerous different approaches have been used to model their properties and kinematics, some of which are listed here.

Debris Flows as Mud Flows

- Rheologically based models that apply to mud flows that are treated as a homogeneous liquid (Examples include: Bingham, viscoplastic, Bagnold-type dilatant fluid, thixotropic, etc.)
- The mixture theory of Iverson
- Dam break wave, e.g. Hunt, Chanson et al.
- Roll wave, e.g., Takahashi, Davies

Unsaturated "Rocky" or "Stony" Debris Flows

- Progressive wave
- A type of translating rock dam

Damage Prevention

In order to prevent debris flows reaching property and people, a debris basin may be constructed. Debris basins are designed to protect soil and water resources or to prevent downstream damage. Such constructions are considered to be a last resort because they are expensive to construct and require commitment to annual maintenance.

Earthflow

An earthflow (earth flow) is a downslope viscous flow of fine-grained materials that have been saturated with water, and moves under the pull of gravity. They are an intermediate type of mass wasting that is between downhill creep and mudflow. The types of materials that are susceptible to earthflows are clay, fine sand and silt, and fine-grained pyroclastic material. When the ground materials become saturated with enough water, they will start flowing. Its speed can range from being barely noticeable to rapid movement. The velocity of the flow is dictated by water content: the higher the water content is, the higher the velocity will be. Because of the dependency on water content for the velocity of the flow, it can take minutes or years for the materials to move down the slope.

An earthflow may affect as few as several square metres or up to several hectares.

Rockfall

Rock fall or rock-fall refers to quantities of rock falling freely from a cliff face. A rockfall is a fragment of rock (a block) detached by sliding, toppling, or falling, that falls along a vertical or sub-vertical cliff, proceeds down slope by bouncing and flying along ballistic trajectories or by rolling on talus or debris slopes," (Varnes, 1978). Alternatively, a "rockfall is the natural downward motion of a detached block or series of blocks with a small volume involving free falling, bouncing, rolling, and sliding".

Causal Mechanisms

Favourable geology and climate are the principal causal mechanisms of rockfall, factors that include intact condition of the rock mass, discontinuities within the rockmass, weathering susceptibility, ground and surface water, freeze-thaw, root-wedging, and external stresses. A tree may be blown by the wind, and this causes a pressure at the root level and this loosens rocks and can trigger a fall. The pieces of rock collect at the bottom creating a *talus* or *scree*. Rocks falling from the cliff may dislodge other rocks and serve to create another mass wasting process, for example an avalanche. A cliff that has favourable geology to a rockfall may be said to be incompetent. One that is not favourable to a rockfall, which is better consolidated, may be said to be competent.

Mitigation

Typically, rockfall events are mitigated in one of two ways: either by passive mitigation or active mitigation. Passive mitigation is where only the effects of the rockfall event are mitigated and are generally employed in the deposition or run-out zones, such as through the use of drape nets, rockfall catchment fences, diversion dams, etc. The rockfall still takes place but an attempt is made to control the outcome. In contrast, active mitigation is carried out in the initiation zone and prevents the rockfall event from ever occurring. Some examples of these measures are rock bolting, slope retention systems, shotcrete, etc. Other active measures might be by changing the geographic or climatic characteristics in the initiation zone, e.g. altering slope geometry, dewatering the slope, revegetation, etc.

The picture shows a road, which stops big rocks.

The mode of failure differs from that of a rockslide.

Design guides of passive measures have been proposed by Ritchie (1963), Pierson et al. (2001) and Pantelidis (2010).

Effects on a Tree

Effect on a tree. The tree roots may rotate - rotational energy. The tree may move - translational energy. And lastly deformation -

either elastic or plastic. Dendrochronology can reveal a past impact, with missing tree rings, as the tree rings grow around and close over a gap. If you look microscopically then you see a difference with a callus tissue. A macroscopic section can be used for dating of avalanche and rock fall events.

Assessment of Risk

Simulation – need a lot of data. Rapid Assessment - based on Mean Tree Free Distance. Gsteiger (1993), the probable mean distance between 2 tree impacts in a forest stand. Should not be longer than about 30m and or steeper than 30 degrees. A rock will just not stop and keeps on going if it is.

Deformation of Rock

Within the Earth rocks are continually being subjected to forces that tend to bend them, twist them, or fracture them. When rocks bend, twist or fracture we say that they deform (change shape or size). The forces that cause deformation of rock are referred to as stresses (Force/unit area). So, to understand rock deformation we must first explore these forces or stresses. Stress and StrainStress is a force applied over an area. One type of stress that we are all used to is a uniform stress, called pressure. A uniform stress is a stress wherein the forces act equally from all directions. In the Earth the pressure due to the weight of overlying rocks is a uniform stress, and is sometimes referred to as confining stress.

If stress is not equal from all directions then we say that the stress is a differential stress. Three kinds of differential stress occur.

1. Tensional stress (or extensional stress), which stretches rock;
2. Compressional stress, which squeezes rock; and
3. Shear stress, which result in slippage and translation.

We can divide materials into two classes that depend on their relative behaviour under stress.

- Brittle materials have a small or large region of elastic behaviour but only a small region of ductile behaviour before they fracture.
- Ductile materials have a small region of elastic behaviour and a large region of ductile behaviour before they fracture.

How a material behaves will depend on several factors. Among them are:

- Temperature - At high temperature molecules and their bonds can stretch and move, thus materials will behave in more ductile manner. At low Temperature, materials are brittle.
- Confining Pressure - At high confining pressure materials are less likely to fracture because the pressure of the surroundings tends to hinder the formation of fractures. At low confining stress, material will be brittle and tend to fracture sooner.
- Strain rate — At high strain rates material tends to fracture. At low strain rates more time is available for individual atoms to move and therefore ductile behaviour is favoured.
- Composition — Some minerals, like quartz, olivine, and feldspars are very brittle. Others, like clay minerals, micas, and calcite are more ductile This is due to the chemical bond types that hold them together. Thus, the mineralogical composition of the rock will be a factor in determining the deformational behaviour of the rock. Another aspect is presence or absence of water. Water appears to weaken the chemical bonds and forms films around mineral grains along which slippage can take place. Thus wet rock tends to behave in ductile manner, while dry rocks tend to behave in brittle manner.

Brittle-Ductile Properties of the Lithosphere

We all know that rocks near the surface of the Earth behave in a brittle manner. Crustal rocks are composed of minerals like quartz and feldspar which have high strength, particularly at low pressure and temperature.

As we go deeper in the Earth the strength of these rocks initially increases. At a depth of about 15 km we reach a point called the brittle-ductile transition zone. Below this point rock strength decreases because fractures become closed and the temperature is higher, making the rocks behave in a ductile manner. At the base of the crust the rock type changes to peridotite which is rich in olivine.

Olivine is stronger than the minerals that make up most crustal rocks, so the upper part of the mantle is again strong. But, just as in the crust, increasing temperature eventually predominates and at a depth of about 40 km the brittle-ductile transition zone in the mantle occurs. Below this point rocks behave in an increasingly ductile manner.

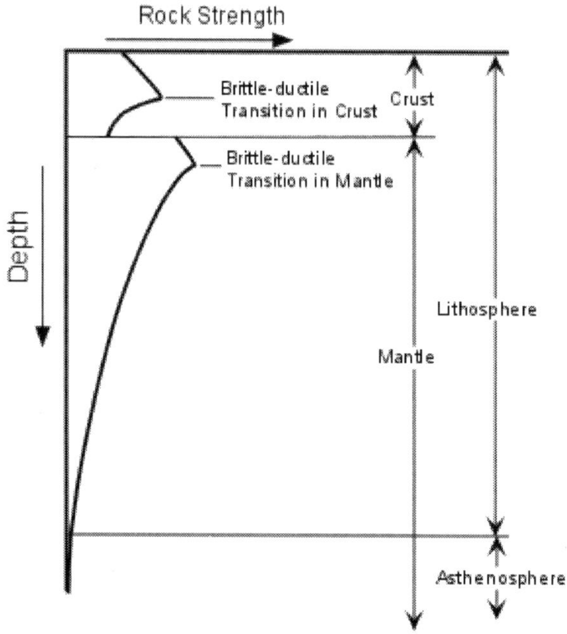

Deformation in Progress

Only in a few cases does deformation of rocks occur at a rate that is observable on human time scales. Abrupt deformation along faults, usually associated with earthquakes caused by the fracture of rocks occurs on a time scale of minutes or seconds. Gradual deformation along faults or in areas of uplift or subsidence can be measured over periods of months to years with sensitive measuring instruments.

Evidence of Former Deformation

Evidence of deformation that has occurred in the past is very evident in crustal rocks. For example, sedimentary strata and lava flows generally follow the law of original horizontality. Thus, when we see such strata inclined instead of horizontal, evidence of an episode of deformation. In order to uniquely define the orientation of a planar feature we first need to define two terms - strike and dip.

For an inclined plane the *strike* is the compass direction of any horizontal line on the plane. The *dip* is the angle between a horizontal plane and the inclined plane, measured perpendicular to the direction of strike. In recording strike and dip measurements on a geologic map, a symbol is used that has a long line oriented parallel to the compass direction of the strike. A short tick mark is placed in the centre of the line on the side to which the inclined plane dips, and the angle of dip is recorded next to the strike and dip symbol as shown above.

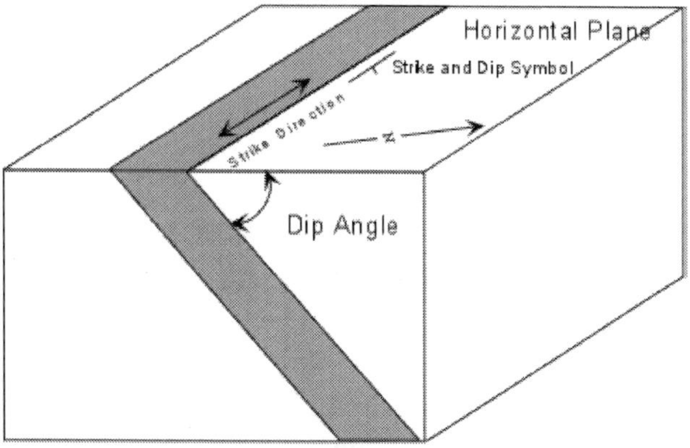

For beds with a 90^0 dip (vertical) the short line crosses the strike line, and for beds with no dip (horizontal) a circle with a cross inside is used as shown below..

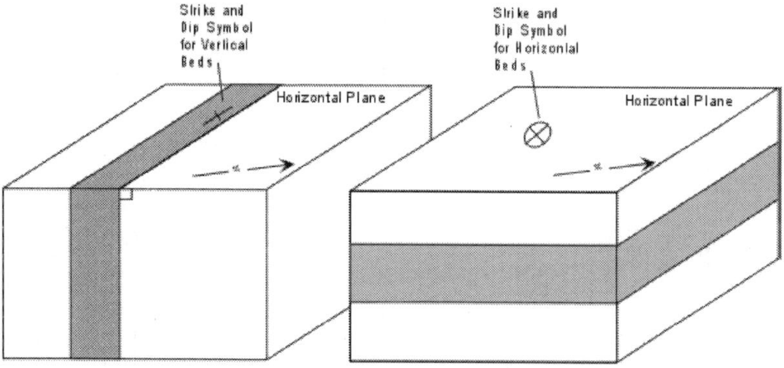

Fracture of Brittle Rocks

Faults - Faults occur when brittle rocks fracture and there is an offset along the fracture. When the offset is small, the displacement can be easily measured, but sometimes the displacement is so large that it is difficult to measure.

Types of Faults

Faults can be divided into several different types depending on the direction of relative displacement. Since faults are planar features, the concept of strike and dip also applies, and thus the strike and dip of a fault plane can be measured. One division of faults is between dip-slip faults, where the displacement is measured along the dip direction of the fault, and strike-slip faults where the displacement is horizontal, parallel to the strike of the fault.

- D*ip Slip Faults:* Dip slip faults are faults that have an inclined fault plane and along which the relative displacement or offset has occurred along the dip direction. Note that in looking at the displacement on any fault we don't know which side actually moved or if both sides moved, all we can determine is the relative sense of motion.

For any inclined fault plane we define the block above the fault as the *hanging wall block* and the block below the fault as the *footwall block.*

- *Normal Faults:* Are faults that result from horizontal tensional stresses in brittle rocks and where the hanging-wall block has moved down relative to the footwall block.

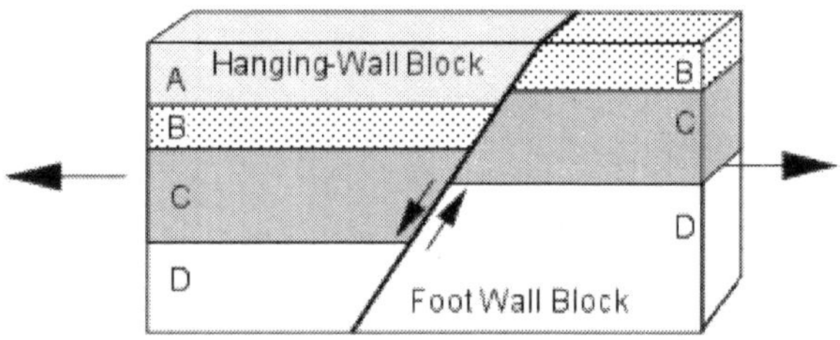

Normal Fault
Extensional Stress

Horsts and Gabens: Due to the tensional stress responsible for normal faults, they often occur in a series, with adjacent faults dipping in opposite directions. In such a case the down-dropped blocks form *grabens* and the uplifted blocks form *horsts.* In areas where tensional stress has recently affected the crust, the grabens may form *rift valleys* and the uplifted horst blocks may form linear mountain ranges. The East African Rift Valley is an example of an area where continental extension has created such a rift. The basin and range province of the western U.S. (Nevada, Utah, and Idaho) is also an area that has recently undergone crustal extension. In the basin and range, the basins are elongated grabens that now form valleys, and the ranges are uplifted horst blocks.

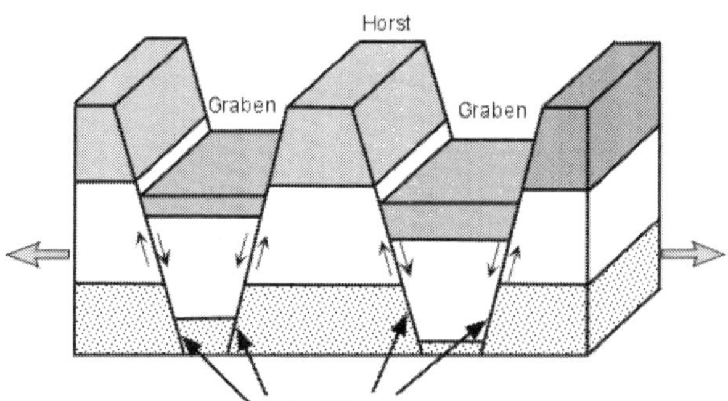

Half-Grabens - A normal fault that has a curved fault plane with the dip decreasing with depth can cause the down-dropped block to rotate. In such a case a half-graben is produced, called such because it is bounded by only one fault instead of the two that form a normal graben.

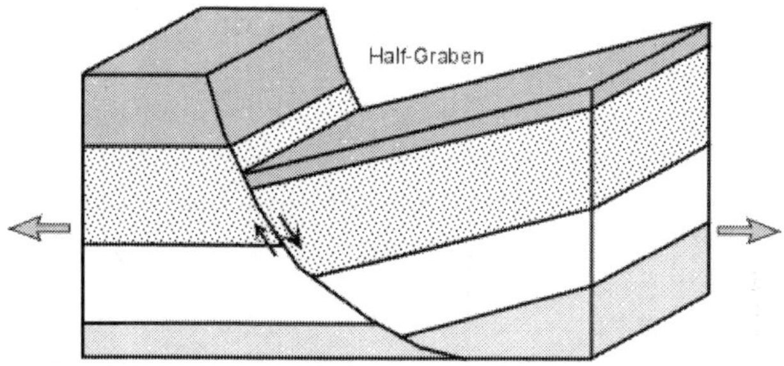

- *Reverse Faults:* are faults that result from horizontal compressional stresses in brittle rocks, where the hanging-wall block has moved up relative the footwall block.

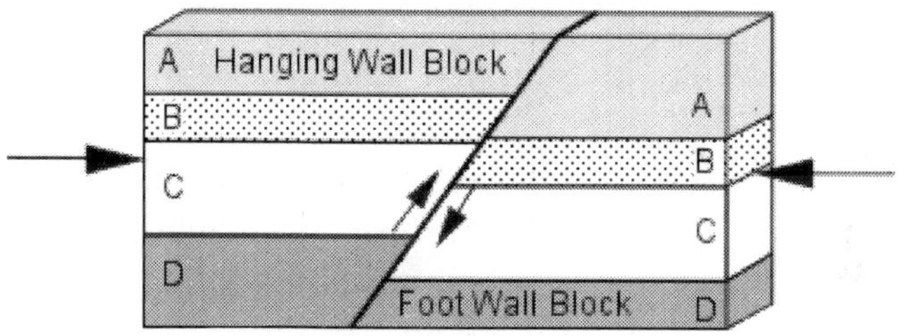

Reverse Fault
Compressional Stress

A *Thrust Fault* is a special case of a reverse fault where the dip of the fault is less than 15°. Thrust faults can have considerable displacement, measuring hundreds of kilometres, and can result in older strata overlying younger strata.

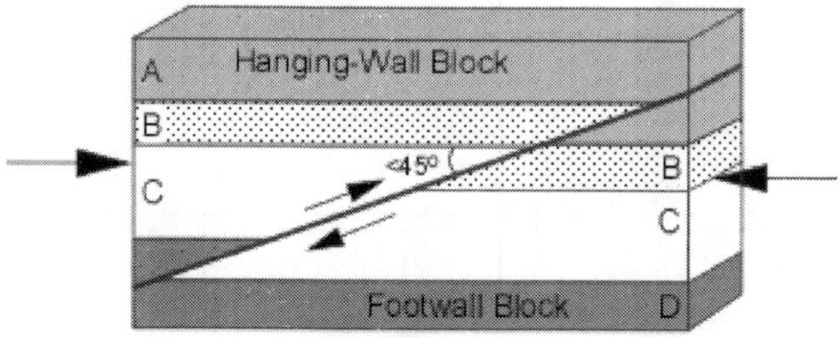

Thrust Fault
Compressional Stress

- *Strike Slip Faults* - are faults where the relative motion on the fault has taken place along a horizontal direction. Such faults result from shear stresses acting in the crust. Strike slip faults can be of two varieties, depending on the sense of displacement. To an observer standing on one side of the fault and looking across the fault, if the block on the other side has moved to the left, we say that the fault is a *left-lateral strike-slip fault.* If the block on the other side has moved to the right, we say that the fault is a *right-lateral strike-slip fault.* The famous San Andreas Fault in California is an example of a right-lateral strike-slip fault. Displacements on the San Andreas fault are estimated at over 600 km.

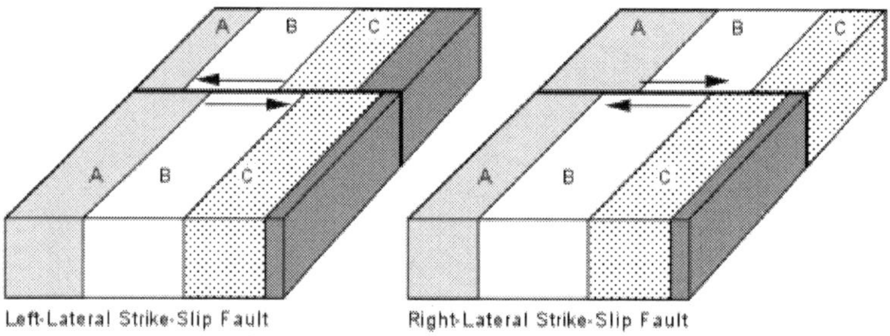

Shear Stress

Transform-Faults are a special class of strike-slip faults. These are plate boundaries along which two plates slide past one another in a horizontal manner. The most common type of transform faults occur where oceanic ridges are offset. Note that the transform fault only occurs between the two segments of the ridge. Outside of this area there is no relative movement because blocks are moving in the same direction. These areas are called fracture zones. The San Andreas fault in California is also a transform fault.

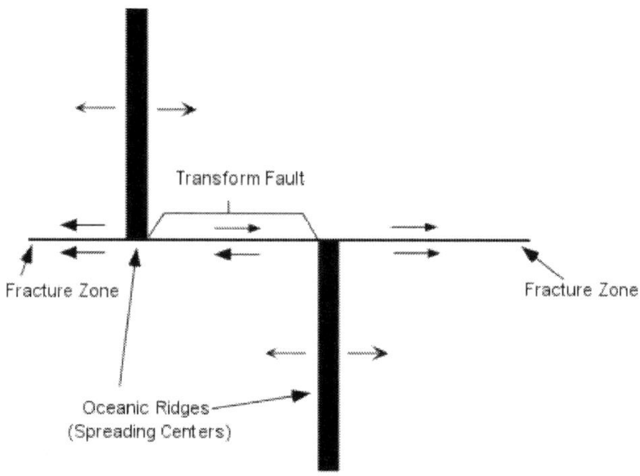

Evidence of Movement on Faults

- *Slikensides* are scratch marks that are left on the fault plane as one block moves relative to the other. Slickensides can be used to determine the direction and sense of motion on a fault.
- *Fault Breccias* are crumbled up rocks consisting of angular fragments that were formed as a result of grinding and crushing movement along a fault.

Folding of Ductile Rocks

When rocks deform in a ductile manner, instead of fracturing to form faults, they may bend or fold, and the resulting structures are called *folds*. Folds result from compressional stresses acting over considerable time. Because the strain rate is low, rocks that we normally consider brittle can behave in a ductile manner resulting in such folds. We recognize several different kinds of folds.

Monoclines are the simplest types of folds. Monoclines occur when horizontal strata are bent upward so that the two limbs of the fold are still horizontal.

Monocline

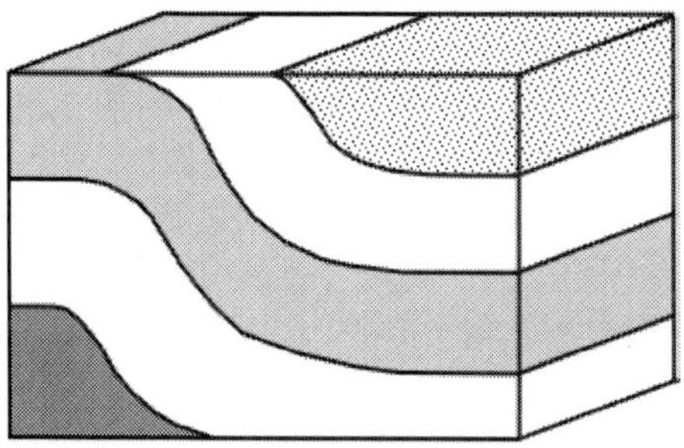

Anticlines are folds where the originally horizontal strata has been folded upward, and the two limbs of the fold dip away from the hinge of the fold.

Anticline

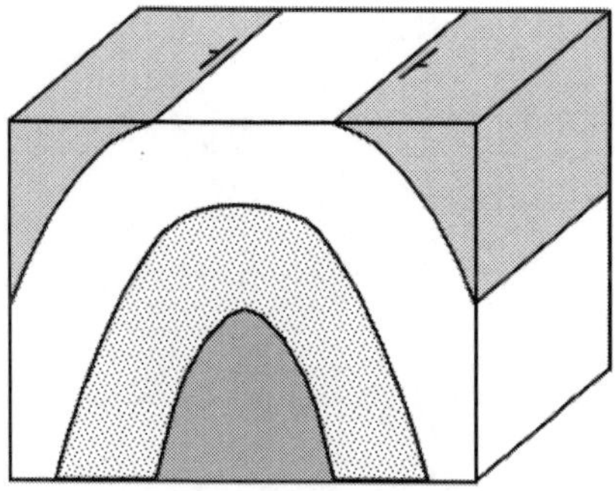

Synclines are folds where the originally horizontal strata have been folded downward, and the two limbs of the fold dip inward towards the hinge of the fold. Synclines and anticlines usually occur together such that the limb of a syncline is also the limb of an anticline.

Syncline

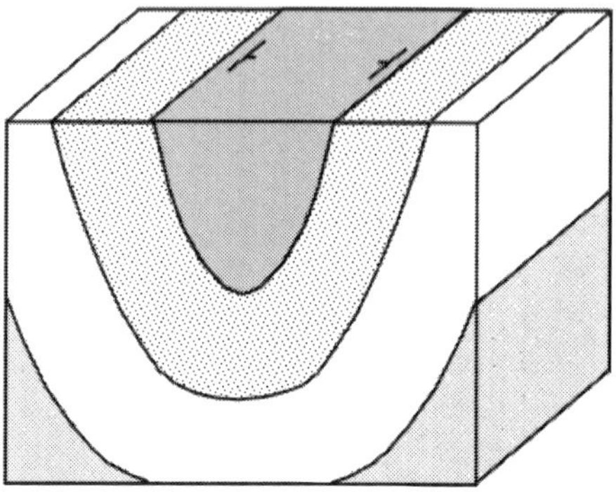

- Geometry of Folds - Folds are described by their form and orientation. The sides of a fold are called *limbs*. The limbs intersect at the tightest part of the fold, called the *hinge*. A line connecting all points on the hinge is called the *fold axis*. In the diagrams above, the fold axes are horizontal, but if the fold axis is not horizontal the fold is called a *plunging fold* and the angle that the fold axis makes with a horizontal line is called the *plunge* of the fold. An imaginary plane that includes the fold axis and divides the fold as symmetrically as possible is called the *axial plane* of the fold.

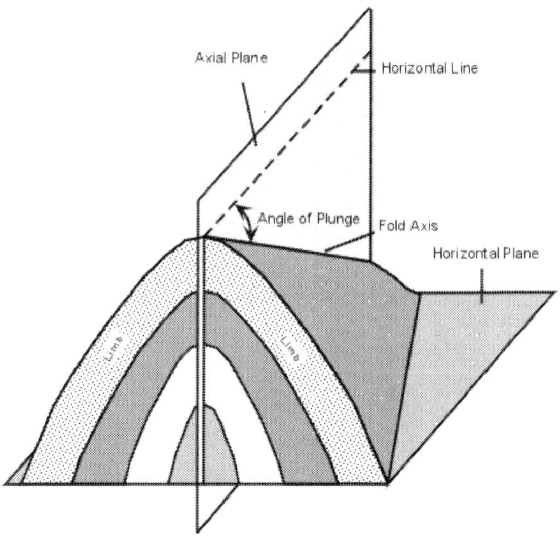

Note that if a plunging fold intersects a horizontal surface, we will see the pattern of the fold on the surface.

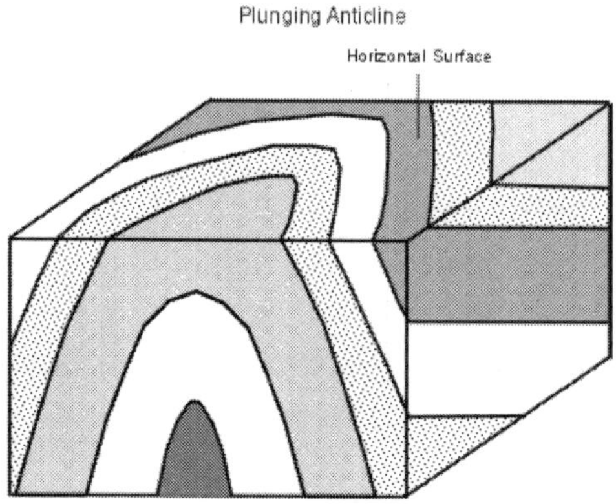

Classification of Folds

- Folds can be classified based on their appearance.
  - o If the two limbs of the fold dip away from the axis with the same angle, the fold is said to be a *symmetrical fold.*
  - o If the limbs dip at different angles, the folds are said to be *asymmetrical folds.*

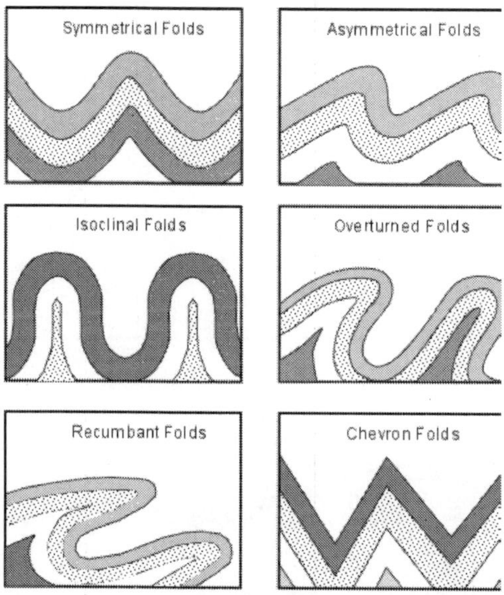

  - o If the compressional stresses that cause the folding are intense, the fold can close up and have limbs that are parallel to each other. Such a fold is called an *isoclinal fold* (iso means same, and cline means angle, so isoclinal means the limbs have the same angle). Note the isoclinal fold depicted in the diagram below is also a symmetrical fold.

- o If the folding is so intense that the strata on one limb of the fold becomes nearly upside down, the fold is called an *overturned fold.*
- o An overturned fold with an axial plane that is nearly horizontal is called a *recumbant fold.*
- o A fold that has no curvature in its hinge and straight-sided limbs that form a zigzag pattern is called a *chevronfold.*

- *The Relationship Between Folding and Faulting:* Because different rocks behave differently under stress, we expect that some rocks when subjected to the same stress will fracture or fault, while others will fold. When such contrasting rocks occur in the same area, such as ductile rocks overlying brittle rocks, the brittle rocks may fault and the ductile rocks may bend or fold over the fault.

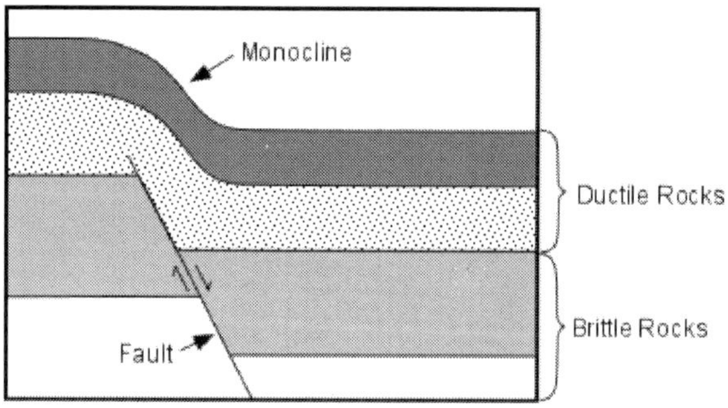

Also since even ductile rocks can eventually fracture under high stress, rocks may fold up to a certain point then fracture to form a fault.

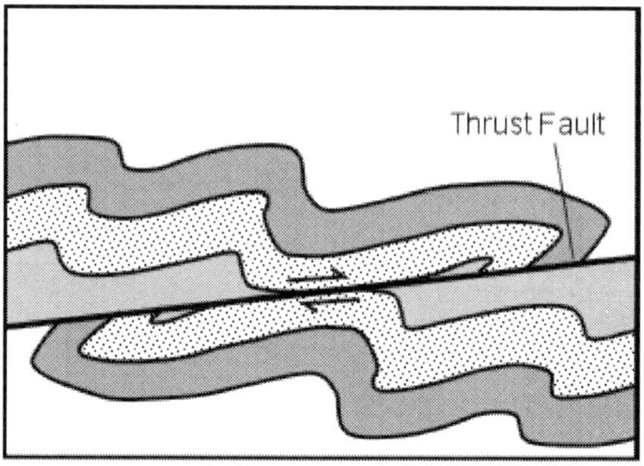

Folds and Topography

Since different rocks have different resistance to erosion and weathering, erosion of folded areas can lead to a topography that reflects the folding. Resistant strata would form ridges that have the same form as the folds, while less resistant strata will form valleys.

Mountain Ranges - The Result of Deformation of the Crust

One of the most spectacular results of deformation acting within the crust of the Earth is the formation of mountain ranges. Mountains originate by three processes, two of which are directly related to deformation. Thus, there are three types of mountains:

1. *Fault Block Mountains* - As the name implies, fault block mountains originate by faulting. As discussed previously, both normal and reverse faults can cause the uplift of blocks of crustal rocks. The Sierra Nevada mountains of California, and the mountains in the Basin and Range province of the western U.S., as discussed previously, were formed by faulting processes and are thus fault block mountains.
2. *Fold & Thrust Mountains* - Large compressional stresses can be generated in the crust by tectonic forces that cause continental crustal areas to collide. When this occurs the rocks between the two continental blocks become folded and faulted under compressional stresses and are pushed upward to form fold and thrust mountains. The Himalayan Mountains (currently the highest on Earth) are mountains of this type and were formed as a result of the Indian Plate colliding with the Eurasian plate. Similarly the Appalachian Mountains of North America and the Alps of Europe were formed by such processes.
3. *Volcanic Mountains* - The third type of mountains, volcanic mountains, are not formed by deformational processes, but instead by the outpouring of magma onto the surface of the Earth. The Cascade Mountains of the western U.S., and of course the mountains of the Hawaiian Islands and Iceland are volcanic mountains.

6

Volcano Landslides and their Effects

Landslides are large masses of rock and soil that fall, slide, or flow very rapidly under the force of gravity. These mixtures of debris move in a wet or dry state, or both. Landslides commonly originate as massive rockslides or avalanches which disintegrate during movement into fragments ranging in size from small particles to enormous blocks hundreds of metres across. If the moving rock debris is large enough and contains a large content of water and fine material (typically, >3-5 percent of clay-sized particles), the landslide may transform into a lahar and flow downvalley more than 100 km from a volcano!

Volcano landslides range in size from less than 1 km3 to more than 100 km3. The high velocity (>100 km/hr) and great momentum of landslides allows them to run up slopes and to cross valley divides up to several hundred metres high. For example, the landslide at Mount St. Helens on May 18, 1980, had a volume of 2.5 km^3, reached speeds of 50-80 m/s (180-288 km/hr), and surged up and over a 400 m tall ridge located about 5 km from the volcano!

Landslides are common on volcanoes because their massive cones (1) typically rise hundreds to thousands of metres above the surrounding terrain; and (2) are often weakened by the very process that created them—the rise and eruption of molten rock. Each time magma moves towards the surface, overlying rocks are shouldered aside as the molten rock makes room for itself, often creating internal shear zones or oversteepening one or more sides of the cone. Magma that remains within the cone releases volcanic gases that partially dissolve in groundwater, resulting in a hot acidic hydrothermal system that weakens rock by altering rock minerals to clay. Furthermore, the tremendous mass of thousands of layers lava and loose fragmented rock debris can lead to internal faults and fault zones that move frequently as the cone "settles" under the downward pull of gravity.

These conditions permit a number of factors to trigger a landslide or to allow part of a volcano's cone to simply collapse under the influence of gravity:

- intrusion of magma into a volcano
- explosive eruptions (magmatic or phreatic—steam-driven explosions)
- large earthquake directly beneath a volcano or nearby (typically >M5)
- intense rainfall that saturates a volcano or adjacent tephra-covered hillslopes with water, especially before or during a large earthquake.

A landslide typically destroys everything in its path and may generate a variety of related activity. Historically, landslides have caused explosive eruptions, buried river valleys with tens of metres of rock debris, generated lahars, triggered waves and tsunami, and created deep horseshoe-shaped craters.

By removing a large part of a volcano's cone, a landslide may abruptly decrease pressure on the shallow magmatic and hydrothermal systems, which can generate explosions ranging from a small steam explosion to large steam- and magma-driven directed blasts. A large landslide often buries valleys with tens to hundreds of metres of rock debris, forming a chaotic landscape marked by dozens of small hills and closed depressions. If the deposit is thick enough, it may dam tributary streams to form lakes in the subsequent days to months; the lakes may eventually drain catastrophically and generate lahars and floods downstream.

Landslides also generate some of the largest and most deadly lahars, either by transforming directly into a lahar or, after it stops moving, from dewatering of the deposit. Historically, however, the most deadly volcano landslide occurred in 1792 when sliding debris from Mt. Mayuyama near Unzen Volcano in Japan slammed into the Ariaka Sea and generated a wave on the opposite side that killed nearly 15,000 people.

On a volcano, landslides typically carve deep gashes into its cone or create large horseshoe-shaped craters hundreds of metres deep and more than a kilometre in width.

Volcanic Landslides

The illustration shows the landslide (green) and directed blast (red) that occurred during the first few minutes of the eruption of Mount St. Helens in 1980.

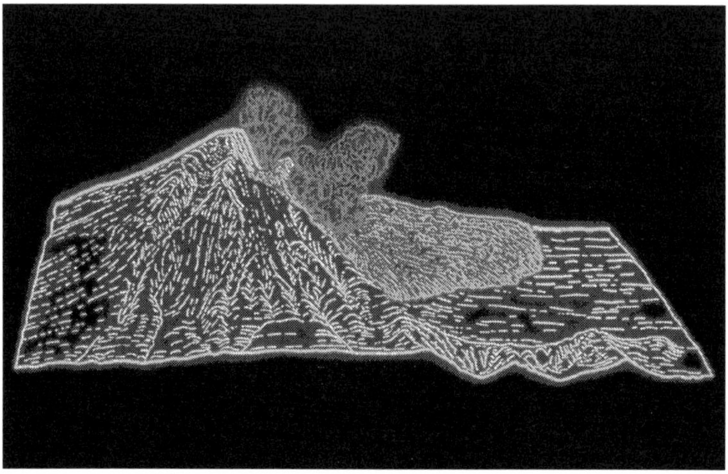

Figure: *trigger volcanic explosions.*

Before the eruption, an estimated 0.11 km^3 of dacite magma had intruded into the volcano (equivalent to sphere about 600 m in diametre!). The rising magma forced the volcano's north flank (right side of illustration) outward about 150 m and heated the volcano's ground water system, causing many steam-driven explosions (phreatic eruptions).

The hot magma and surrounding hydrothermal system were unroofed by the landslide (green), and the resulting rapid depressurization caused a series of steam- and volcanic-gas-driven explosions. The explosions burst through part of the landslide, blasting rock debris northward. The resulting pyroclastic surge quickly overran the landslide and spread over ridges and valleys across an area of 550 km^2.

This house is partially buried in a lahar deposit that was formed by the dewatering of a large volcano landslide from Mount St. Helens, Washington. Early on the morning of May 18, 1980, the landslide swept into the upper North Fork Toutle River valley and came to rest within about 22 km of the volcano. The landslide deposit, however, was saturated with water, and contained snow and ice blocks from the volcano's former glaciers. As soon as the landslide stopped moving, water percolated to the top of the deposit and poured across its irregular surface, forming many lahars that merged as they rushed down the valley. The peak flow swept from the deposit about 5 hours after the landslide was emplaced!

The lahar flowed down the Toutle River throughout the afternoon and evening, reaching its peak at midnight about 60 km downstream from the volcano. The lahar destroyed roads, bridges, and homes.

Many volcano landslides do not stop so close to their source, but instead keep moving by transforming directly into a lahar. These lahars can be extremely hazardous because of their size and mobility (they may travel more than 100 km).

View of Mt. Mayuyama is towards the north (Photograph by T. Casadevall in 1991). Mt. Mayuyama is one of the dacite lava domes that form the Unzen Volcano complex on Kyushu Island, Japan. The east flank of Mt. Mayuyama collapsed without warning on May 21, 1792, and generated a landslide that swept through Shimabara City and slammed into the Ariaka Sea. The displacement of water triggered a tsunami along the adjacent shoreline of Shimabara Peninsula (visible above and right of Mayuyama) and also 17-23 km across the Ariaka Sea in neighbouring provinces. The landslide and tsunami killed nearly 15,000 people, Japan's worst historical volcanic disaster.

Scientists have interpreted the conspicuous hummocks along the shore as part of a landslide deposit that occurred before 1792. Maps submitted to the Tokugawa Shogunate in 1792 as the official documents of the Shimabara Catastrophe clearly show the existence of small islands before the disaster struck.

Other historical volcano landslides are known to have generated tsunami, including:

- landslide from Kamagatake volcano on Hokkaido Island, Japan, in 1640 killed 700 people
- landslide from Oshima-Oshima Volcano on Hokkaido Island, Japan, in 1741-42 killed 1,474 people on Hokkaido and northern Honshu
- landslide from Augustine Volcano, Alaska, in 1883 triggered a tsunami that swept across Cook Inlet onto the Kenai Peninsula but caused no damage

A scientist stands on one of the many small hills called hummocks that form the chaotic surface of a massive landslide deposit in the upper North Fork Toutle River valley below Mount St. Helens volcano (10 km in distance). Before the landslide and eruption on May 18, 1980, a forest grew on this part of the valley floor, and a highway followed the meandering river to Spirit Lake, a popular recreation area.

The landslide deposit extends about 22 km from the volcano and buries the river valley to an average depth of about 45 m. In places, the deposit is nearly 200 m thick! The landslide covers an area of about 60 km^2.

An exceptionally large landslide deposit was discovered at Mount Shasta shortly after the eruption of Mount St. Helens. This landslide has a volume of about 45 km^3—nearly 20 times larger than the one that buries the North Fork Toutle River valley (above)—and it covers an area of 675 km^2. Photograph by L. Topinka in 1981.

Figure: *dam tributary streams to form lakes.*

View is looking northwest up the valley of former Coldwater Creek, now filled with a lake. When the landslide from Mount St. Helens slid into the North Fork Toutle River valley (foreground), it blocked the flow of Coldwater Creek. Water backed up behind the landslide deposit, gradually forming a lake about 8 km long and 55 m deep. The landslide was rushed down the Toutle valley from right to left.

Concern about the sudden breakout of water from Coldwater Lake from failure of the drebris dam or overtopping and subsequent erosion of the dam, led the Corps of Engineers in 1981 to control the lake level by excavating an outlet channel that delivers water to the Toutle River.

The fan-shaped delta on the southeast shore of Coldwater Lake forms where a stream from South Fork Coldwater Creek pours into the lake. The delta began to grow quickly when water from Spirit

Lake was diverted into Coldwater Creek beginning in 1985. A long tunnel was drilled through a ridge to deliver water from Spirit Lake into South Fork Coldwater Creek in order to stabilize the level of Spirit Lake. Photograph by L. Topinka on January 13, 1984.

Figure: *create a crater or scar on volcano.*

View is looking south into the crater of Mount St. Helens formed by an enormous landslide on May 18, 1980. The newly-formed crater is about 2 km wide (east-west), 3 km long (north-south), and about 600 m deep. The landslide removed about 2.3 km^3 from the volcano's cone, which towered 1,035 m above the crater floor!

Large horseshoe-shaped craters, open at one end, have long been noted in many volcanic regions around the world. The origin of these breached craters has been controversial, but since the landslide and eruption of Mount St. Helens in 1980, many have been interpreted by scientists as the result of a landslide.

If a large landslide creates a horseshoe-shaped crater that exposes a volcano's eruptive vent, the deep crater will likely direct subsequent volcanic activity (lava flows, pyroclastic flows, or lahars) towards its breached opening. A new hazard assessment may be necessary to determine the way in which volcano-hazard areas downslope from the crater may have changed. Photograph by C.D. Miller in 1980.

Powder Snow Avalanche

A powder snow avalanche is a type of avalanche where the snow grains are largely or wholly suspended by fluid turbulence. They are particle-laden gravity currents and closely related to turbidity currents, pyroclastic flows from volcanoes and dust storms in the desert. The turbulence is typically generated by the forward motion of the current along the lower boundary of the domain, the motion being in turn driven by the action of gravity on the density difference between the particle-fluid mixture and the ambient fluid. The ambient fluid is generally of similar composition to (and miscible with) the interstitial fluid, and is water for turbidity currents and air for avalanches. These flows are non-conservative in that they may exchange particles at the lower boundary by deposition or suspension, and may exchange fluid with the ambient by entrainment or detrainment. Such flows dissipate when the turbulence can no longer hold the particles in suspension and they are deposited on the lower boundary. When the turbulence is strong enough to suspend new material from the bed or the underlying dense flow then current is said to be auto-suspending. Particle concentrations in the suspension cloud are usually sufficiently low (0.1—7\,\% by volume) that particle-particle interactions play a small or negligible role in maintaining the suspension. In powder snow avalanches, even at these low concentrations, the extra density of the suspended particles is large relative to that of water, so the Boussinesq approximation, where density differences are considered negligible in inertia terms, is invalid, so that the snow grains carry most of the flows momentum. This is in contrast to turbidity currents and laboratory experiments in water where the extra inertia of the particles can usually be neglected. Nonetheless, due to the extreme difficulty in estimating particle concentrations in natural flows there remains considerable uncertainty — and debate — concerning the particle loading in large submarine turbidity currents and the validity of the Boussinesq approximation.

Loose Snow Avalanche

A loose snow avalanche is an avalanche formed in snow with little internal cohesion among individual snow crystals. Usually very few fatalities occur from loose snow avalanches, as the avalanches have a tendency to break beneath the person and are usually small even having a path as small as a few centimetres, and as a result are sometimes called "harmless sloughs" that usually at most cause the person to merely fall. However based on the terrain loose snow avalanches can grow large, and have been known to carry people off a cliff and into a crevass or bury them in a gully, and even completely destroy houses and other similar buildings. Ideal conditions for a loose snow avalanche are steep slope angles of 40 degrees and more, persistent sub-zero temperatures and low humidity, moderate to heavy snowfall, and also in an area where winds are very light or are not affecting the density of the snow. This produces light, fluffy snow that is hard or unable to pack. Small loose snow avalanches can even be a sign of stability of the snow, as slabs are triggered by a hard layer of snow over a very soft and weak layer of snow, and loose snow avalanches consist of a very soft layer of snow on a hard layer or the ground.

Pyroclastic Flow

A pyroclastic flow (also known scientifically as a pyroclastic density current) is a fast-moving current of hot gas and rock (collectively known as tephra), which reaches speeds moving away from a volcano of up to 700 km/h (450 mph). The gas can reach temperatures of about 1,000 °C (1,830 °F). Pyroclastic flows normally hug the ground and travel downhill, or spread laterally under gravity. Their speed depends upon the density of the current, the volcanic output rate, and the gradient of the slope. They are a common and devastating result of certain explosive volcanic eruptions.

Origin of Term

The word *pyroclast* is derived from the Greekðæñ, meaning "fire", and êëáóôüò, meaning "broken in pieces". A name for some pyroclastic flows is *nuée ardente* (French for "burning cloud"); this was first used to describe the disastrous 1902 eruption of Mount Pelée on Martinique. In the dark, these pyroclastic flows glowed red.

Pyroclastic flows that contain a much higher proportion of gas to rock are known as "fully dilute pyroclastic density currents" or pyroclastic surges. The lower density sometimes allows them to flow over higher topographic features such as ridges and hills. They may also contain steam, water and rock at less than 250 °C (482 °F); these are called "cold" compared with other flows, although the temperature is still lethally high. Cold pyroclastic surges can occur when the eruption is from a vent under a shallow lake or the sea. Fronts of some pyroclastic density currents are fully dilute; for example, during the eruption of Mount Pelée in 1902 a fully dilute current overwhelmed the city of Saint-Pierre and killed nearly 30,000 people.

A pyroclastic flow is a type of gravity current; in scientific literature they are sometimes abbreviated to PDC (pyroclastic density current).

Causes

There are several factors which can produce a pyroclastic flow:

- *Fountain collapse* of an eruption column from a Plinian eruption (e.g., Mount Vesuvius's destruction of Pompeii). In such an eruption, the material ejected from the vent heats the surrounding air and the turbulent mixture rises, through convection, for many kilometres. If the erupted jet is unable to heat the surrounding air sufficiently, convection currents will not be strong enough to carry the plume upwards and it falls, flowing down the flanks of the volcano.

- *Fountain collapse* of an eruption column associated with a vulcanian eruption (e.g., Montserrat's Soufrière Hills volcano has generated many of these deadly pyroclastic flows and surges.) The gas and projectiles create a cloud which is denser than the surrounding air and becomes a pyroclastic flow.
- Frothing at the mouth of the vent during degassing of the erupted lava. This can lead to the production of a rock called ignimbrite. This occurred during the eruption of Novarupta in 1912.
- Gravitational collapse of a lava dome or spine, with subsequent avalanches and flow down a steep slope (e.g., Montserrat's Soufrière Hills volcano, which caused nineteen deaths in 1997).
- The directional blast (or jet) when part of a volcano explodes or collapses (e.g., the May 18, 1980, eruption of Mount St. Helens). As distance from the volcano increases, this rapidly transforms into a gravity-driven current.

Size and Effects

The volumes range from a few hundred cubic metres to more than a thousand cubic kilometres. The larger ones can travel for hundreds of kilometres, although none on that scale have occurred for several hundred thousand years. Most pyroclastic flows are around one to ten cubic kilometres and travel for several kilometres. Flows usually consist of two parts: the *basal flow* hugs the ground and contains larger, coarse boulders and rock fragments, while an extremely hot ash plume lofts above it because of the turbulence between the flow and the overlying air, admixes and heats cold atmospheric air causing expansion and convection.

The kinetic energy of the moving boulders will flatten trees and buildings in their path. The hot gases and high speed make them particularly lethal, as they will incinerate living organisms instantaneously:

- The towns of Pompeii and Herculaneum, Italy, for example, were engulfed by pyroclastic surges in 79 AD with many lives lost.

- A pyroclastic surge killed volcanologists Katia and Maurice Krafft and 41 other people on Mount Unzen, in Japan, on June 3, 1991. The surge started as a pyroclastic flow and the more energised surge climbed a spur on which the Kraffts and the others were standing; it engulfed them, and the corpses were covered with about 5 mm of ash.
- On 25 June 1997 a pyroclastic flow travelled down Mosquito Ghaut on the Caribbean island of Montserrat. A large energised pyroclastic surge developed. This surge could not be restrained by the Ghaut and spilled out of it, killing 19 people who were in the Streatham village area (which was officially evacuated). Several others in the area suffered severe burns.

Interaction with Water

Testimonial evidence from the 1883 eruption of Krakatoa, supported by experimental evidence, shows that pyroclastic flows can cross significant bodies of water. One flow reached the Sumatran coast as much as 48 km (30 mi) away.

A 2006 documentary film, *Ten Things You Didn't Know About Volcanoes*, demonstrated tests by a research team at Kiel University, Germany, of pyroclastic flows moving over water. When the reconstructed pyroclastic flow (stream of mostly hot ash with varying densities) hit the water two things happened. The heavier material fell into the water, precipitating out from the pyroclastic flow and into the liquid. The temperature of the ash caused the water to evaporate, propelling the pyroclastic flow (now only consisting of the lighter material) along at an even faster pace than before on a bed of steam.

During some phases of the Soufriere Hills volcano on Montserrat, pyroclastic flows were filmed about 1 km offshore. These show the water boiling as the flow passed over it. The flows eventually built a delta which covered about 1 km^2.

A pyroclastic flow can interact with a body of water to form a large amount of mud, which can then continue to flow downhill as a lahar. This is one of several mechanisms that can create a lahar.

On the Moon

In 1963, NASA astronomer Winifred Cameron proposed that sinuous rilles on the Moon may have been formed by the lunar equivalent of terrestrial pyroclastic flows. In a lunar volcanic eruption, a pyroclastic cloud would follow local relief resulting in an often sinuous track. The Moon's Schroter's Valley is one example.

Volcano

A volcano is an opening, or rupture, in the surface or crust of the Earth or a planetary mass object, which allows hot lava, volcanic ash and gases to escape from the magma chamber below the surface.

On Earth, volcanoes are generally found where tectonic plates are diverging or converging. A mid-oceanic ridge, for example the Mid-Atlantic Ridge, has examples of volcanoes caused by divergent tectonic plates pulling apart; the Pacific Ring of Fire has examples of volcanoes caused by convergent tectonic plates coming together. By contrast, volcanoes are not usually created where two tectonic plates slide past one another. Volcanoes can also form where there is stretching and thinning of the Earth's crust in the interiors of plates, e.g., in the East African Rift, the Wells Gray-Clearwater volcanic field and the Rio Grande Rift in North America. This type of volcanism falls under the umbrella of "Plate hypothesis" volcanism. Volcanism away from plate boundaries has also been explained as mantle plumes. These so-called "hotspots", for example Hawaii, are postulated to arise from upwelling diapirs with magma from the core–mantle boundary, 3,000 km deep in the Earth.

Erupting volcanoes can pose many hazards, not only in the immediate vicinity of the eruption. Volcanic ash can be a threat to aircraft, in particular those with jet engines where ash particles can be melted by the high operating temperature; the melted particles then adhere to the turbine blades and alter their shape, disrupting the operation of the turbine. Large eruptions can affect temperature as ash and droplets of sulfuric acid obscure the sun and cool the Earth's lower atmosphere or troposphere; however, they also absorb heat radiated up from the Earth, thereby warming the stratosphere. Historically, so-called volcanic winters have caused catastrophic famines.

Etymology

The word *volcano* is derived from the name of Vulcano, a volcanic island in the Aeolian Islands of Italy whose name in turn originates from Vulcan, the name of a god of fire in Roman mythology. The study of volcanoes is called volcanology, sometimes spelled *vulcanology*.

Plate Tectonics

Divergent Plate Boundaries: At the mid-oceanic ridges, two tectonic plates diverge from one another. New oceanic crust is being formed by hot molten rock slowly cooling and solidifying. The crust is very thin at mid-oceanic ridges due to the pull of the tectonic plates.

The release of pressure due to the thinning of the crust leads to adiabatic expansion, and the partial melting of the mantle causing volcanism and creating new oceanic crust. Most divergent plate boundaries are at the bottom of the oceans, therefore most volcanic activity is submarine, forming new seafloor. Black smokers or deep sea vents are an example of this kind of volcanic activity. Where the mid-oceanic ridge is above sea-level, volcanic islands are formed, for example, Iceland.

Convergent Plate Boundaries

Subduction zones are places where two plates, usually an oceanic plate and a continental plate, collide. In this case, the oceanic plate subducts, or submerges under the continental plate forming a deep ocean trench just offshore. In a process called flux melting, water released from the subducting plate lowers the melting temperature of the overlying mantle wedge, creating magma. This magma tends to be very viscous due to its high silica content, so often does not reach the surface and cools at depth. When it does reach the surface, a volcano is formed. Typical examples for this kind of volcano are Mount Etna and the volcanoes in the Pacific Ring of Fire.

"Hotspots"

"Hotspots" is the name given to volcanic provinces postulated to be formed by mantle plumes. These are postulated to comprise columns of hot material that rise from the core-mantle boundary. They are suggested to be hot, causing large-volume melting, and to be fixed in space. Because the tectonic plates move across them, each volcano becomes dormant after a while and a new volcano is then formed as the plate shifts over the postulated plume. The Hawaiian Islands have been suggested to have been formed in such a manner, as well as the Snake River Plain, with the Yellowstone Caldera being the part of the North American plate currently above the hot spot. This theory is currently under criticism, however.

Volcanic Features

The most common perception of a volcano is of a conical mountain, spewing lava and poisonous gases from a crater at its summit. This describes just one of many types of volcano, and the features of volcanoes are much more complicated. The structure and behaviour of volcanoes depends on a number of factors. Some volcanoes have rugged peaks formed by lava domes rather than a summit crater, whereas others present landscape features such as massive plateaus.

Figure: *Lakagigar fissure vent in Iceland, source of the major world climate alteration of 1783–84.*

Figure: *Skjaldbreiður, a shield volcano whose name means "broad shield"*

Vents that issue volcanic material (lava, which is what magma is called once it has escaped to the surface, and ash) and gases (mainly steam and magmatic gases) can be located anywhere on the landform. Many of these vents give rise to smaller cones such as on a flank of Hawaii's Kîlauea. Other types of volcano include cryovolcanoes (or ice volcanoes), particularly on some moons of Jupiter, Saturn and Neptune; and mud volcanoes, which are formations often not associated with known magmatic activity. Active mud volcanoes tend to involve temperatures much lower than those of igneous volcanoes, except when a mud volcano is actually a vent of an igneous volcano.

Fissure Vents

Volcanic fissure vents are flat, linear cracks through which lava emerges.

Shield Volcanoes

Shield volcanoes, so named for their broad, shield-like profiles, are formed by the eruption of low-viscosity lava that can flow a great distance from a vent. They generally do not explode catastrophically. Since low-viscosity magma is typically low in silica, shield volcanoes are more common in oceanic than continental settings. The Hawaiian volcanic chain is a series of shield cones, and they are common in Iceland, as well.

Lava Domes

Lava domes are built by slow eruptions of highly viscous lavas. They are sometimes formed within the crater of a previous volcanic eruption (as in Mount Saint Helens), but can also form independently, as in the case of Lassen Peak. Like stratovolcanoes, they can produce violent, explosive eruptions, but their lavas generally do not flow far from the originating vent.

Cryptodomes

Cryptodomes are formed when viscous lava forces its way up and causes a bulge. The 1980 eruption of Mount St. Helens was an example. Lava was under great pressure and forced a bulge in the mountain, which was unstable and slid down the north side.

Volcanic Cones (Cinder Cones)

Volcanic cones or cinder cones result from eruptions of mostly small pieces of scoria and pyroclastics (both resemble cinders, hence the name of this volcano type) that build up around the vent. These can be relatively short-lived eruptions that produce a cone-shaped hill perhaps 30 to 400 metres high. Most cinder cones erupt only once. Cinder cones may form as flank vents on larger volcanoes, or occur on their own. Parícutin in Mexico and Sunset Crater in Arizona are examples of cinder cones. In New Mexico, Caja del Rio is a volcanic field of over 60 cinder cones.

Based on satellite images it was suggested that cinder cones might occur on other terrestrial bodies in the Solar system too; on the surface of Mars and Moon.

Stratovolcanoes (Composite Volcanoes)

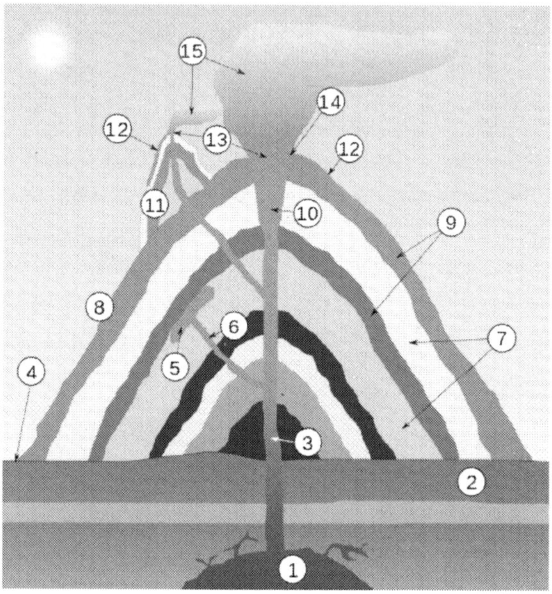

Figure: *Cross-section through a stratovolcano (vertical scale is exaggerated): 1. Large magma chamber; 2. Bedrock; 3. Conduit (pipe); 4. Base; 5. Sill; 6. Dike; 7. Layers of ash emitted by the volcano; 8. Flank; 9. Layers of lava emitted by the volcano; 10. Throat; 11. Parasitic cone; 12. Lava flow; 13. Vent; 14.; Crater; 15. Ash cloud*

Stratovolcanoes or composite volcanoes are tall conical mountains composed of lava flows and other ejecta in alternate layers, the strata that give rise to the name. Stratovolcanoes are also known as composite volcanoes, created from several structures during different kinds of eruptions. Strato/composite volcanoes are made of cinders, ash and lava. Cinders and ash pile on top of each other, lava flows on top of the ash, where it cools and hardens, and then the process begins again. Classic examples include Mt. Fuji in Japan, Mayon Volcano in the Philippines, and Mount Vesuvius and Stromboli in Italy.

Throughout recorded history, ash produced by the explosive eruption of stratovolcanoes has posed the greatest hazard to civilizations as compared to other types of volcanoes. Shield volcanos have smaller pressure buildup from the underlying lava flow as compared to stratovolcanoes. Fissure vents and monogenetic volcanic fields (volcanic cones) have less powerful eruptions, as they are many times under extension. Stratovolcanoes have been a greater historical threat because they are steeper than shield volcanos, with slopes of 30–35° compared to slopes of generally 5–10°, and their loose tephra are material for dangerous lahars.

Supervolcanoes

A supervolcano is a large volcano that usually has a large caldera and can potentially produce devastation on an enormous, sometimes continental, scale. Such eruptions would be able to cause severe cooling of global temperatures for many years afterwards because of the huge volumes of sulfur and ash erupted. They are the most dangerous type of volcano. Examples include Yellowstone Caldera in Yellowstone National Park and Valles Caldera in New Mexico (both western United States), Lake Taupo in New Zealand, Lake Toba in Sumatra, Indonesia and Ngorogoro Crater in Tanzania, Krakatoa near Java and Sumatra, Indonesia. Supervolcanoes are hard to identify centuries later, given the enormous areas they cover. Large igneous provinces are also considered supervolcanoes because of the vast amount of basalt lava erupted, but are non-explosive.

Submarine Volcanoes

Submarine volcanoes are common features on the ocean floor. Some are active and, in shallow water, disclose their presence by blasting steam and rocky debris high above the surface of the sea. Many others lie at such great depths that the tremendous weight of the water above them prevents the explosive release of steam and gases, although they can be detected by hydrophones and discolouration of water because of volcanic gases. Pumice rafts may also appear. Even large submarine eruptions may not disturb the ocean surface. Because of the rapid cooling effect of water as compared to air, and increased buoyancy, submarine volcanoes often form rather steep pillars over their volcanic vents as compared to above-surface volcanoes. They may become so large that they break the ocean surface as new islands. Pillow lava is a common eruptive product of submarine volcanoes. Hydrothermal vents are common near these volcanoes, and some support peculiar ecosystems based on dissolved minerals.

Subglacial Volcanoes

Subglacial volcanoes develop underneath icecaps. They are made up of flat lava which flows at the top of extensive pillow lavas and palagonite. When the icecap melts, the lavas on the top collapse, leaving a flat-topped mountain. These volcanoes are also called table mountains, tuyas or (uncommonly) mobergs. Very good examples of this type of volcano can be seen in Iceland, however, there are also tuyas in British Columbia. The origin of the term comes from Tuya Butte, which is one of the several tuyas in the area of the Tuya River and Tuya Range in northern British Columbia. Tuya Butte was the first such landform analyzed and so its name has entered the geological literature for this kind of volcanic formation. The Tuya Mountains Provincial Park was recently established to protect this unusual landscape, which lies north of Tuya Lake and south of the Jennings River near the boundary with the Yukon Territory.

Mud Volcanoes

Mud volcanoes or mud domes are formations created by geo-excreted liquids and gases, although there are several processes which may cause such activity. The largest structures are 10 kilometres in diametre and reach 700 metres high.

Erupted Material

Figure: *The Stromboli stratovolcano off the coast of Sicily has erupted continuously for thousands of years, giving rise to the term strombolian eruption.*

Lava Composition

Another way of classifying volcanoes is by the *composition of material erupted* (lava), since this affects the shape of the volcano. Lava can be broadly classified into 4 different compositions (Cas & Wright, 1987):

- If the erupted magma contains a high percentage (>63%) of silica, the lava is called felsic.
    - o Felsic lavas (dacites or rhyolites) tend to be highly viscous (not very fluid) and are erupted as domes or short, stubby flows. Viscous lavas tend to form stratovolcanoes or lava domes. Lassen Peak in California is an example of a volcano formed from felsic lava and is actually a large lava dome.
    - o Because siliceous magmas are so viscous, they tend to trap volatiles (gases) that are present, which cause the magma to erupt catastrophically, eventually forming stratovolcanoes. Pyroclastic flows (ignimbrites) are highly hazardous products of such volcanoes, since they are composed of molten volcanic ash too heavy to go up into the atmosphere, so they hug the volcano's slopes and travel far from their vents during large eruptions. Temperatures as high as 1,200 °C are known to occur in pyroclastic flows, which will incinerate everything flammable in their path and thick layers of hot pyroclastic flow deposits can be laid down, often up to many metres thick. Alaska's Valley of Ten Thousand Smokes, formed by the eruption of Novarupta near Katmai in 1912, is an example of a thick pyroclastic flow or ignimbrite deposit. Volcanic ash that is light enough to be erupted high into the Earth's atmosphere may travel many kilometres before it falls back to ground as a tuff.
- If the erupted magma contains 52–63% silica, the lava is of *intermediate* composition.
    - o These "andesitic" volcanoes generally only occur above subduction zones (e.g. Mount Merapi in Indonesia).
    - o Andesitic lava is typically formed at convergent boundary margins of tectonic plates, by several processes:
        - – Hydration melting of peridotite and fractional crystallization
        - – Melting of subducted slab containing sediments
        - – Magma mixing between felsic rhyolitic and mafic basaltic magmas in an intermediate reservoir prior to emplacement or lava flow.
- If the erupted magma contains <52% and >45% silica, the lava is called mafic (because it contains higher percentages of magnesium (Mg) and iron (Fe)) or basaltic. These lavas are

usually much less viscous than rhyolitic lavas, depending on their eruption temperature; they also tend to be hotter than felsic lavas. Mafic lavas occur in a wide range of settings:

- o At mid-ocean ridges, where two oceanic plates are pulling apart, basaltic lava erupts as pillows to fill the gap;
- o Shield volcanoes (e.g. the Hawaiian Islands, including Mauna Loa and Kilauea), on both oceanic and continental crust;
- o As continental flood basalts.

- Some erupted magmas contain <=45% silica and produce ultramafic lava. Ultramafic flows, also known as komatiites, are very rare; indeed, very few have been erupted at the Earth's surface since the Proterozoic, when the planet's heat flow was higher. They are (or were) the hottest lavas, and probably more fluid than common mafic lavas.

Lava Texture

Two types of lava are named according to the surface texture and pâhoehoe, both Hawaiian words. »A»a is characterized by a rough, clinkery surface and is the typical texture of viscous lava flows. However, even basaltic or mafic flows can be erupted as »a»a flows, particularly if the eruption rate is high and the slope is steep.

Pâhoehoe is characterized by its smooth and often ropey or wrinkly surface and is generally formed from more fluid lava flows. Usually, only mafic flows will erupt as pâhoehoe, since they often erupt at higher temperatures or have the proper chemical make-up to allow them to flow with greater fluidity.

Popular Classification of Volcanoes

A popular way of classifying magmatic volcanoes is by their frequency of eruption, with those that erupt regularly called active, those that have erupted in historical times but are now quiet called dormant or inactive, and those that have not erupted in historical times called extinct. However, these popular classifications—extinct in particular—are practically meaningless to scientists. They use classifications which refer to a particular volcano's formative and eruptive processes and resulting shapes, which was explained above.

Active

There is no consensus among volcanologists on how to define an "active" volcano. The lifespan of a volcano can vary from months to several million years, making such a distinction sometimes meaningless when compared to the lifespans of humans or even civilizations. For example, many of Earth's volcanoes have erupted dozens of times in the past few thousand years but are not currently showing signs of eruption. Given the long lifespan of such volcanoes, they are very active. By human lifespans, however, they are not.

Scientists usually consider a volcano to be *erupting* or *likely to erupt* if it is currently erupting, or showing signs of unrest such as unusual earthquake activity or significant new gas emissions. Most scientists consider a volcano *active* if it has erupted in the last 10,000 years (Holocene times) – the Smithsonian Global Volcanism Programme uses this definition of *active*. There are about 1500 active volcanoes in the world – the majority along the Pacific Ring of Fire – and around 50 of these erupt each year. An estimated 500 million people live near active volcanoes.

Historical time (or recorded history) is another timeframe for *active*. The *Catalogue of the Active Volcanoes of the World*, published by the International Association of Volcanology, uses this definition, by which there are more than 500 active volcanoes. However, the span of recorded history differs from region to region. In China and the Mediterranean, it reaches back nearly 3,000 years, but in the Pacific Northwest of the United States and Canada, it reaches back less than 300 years, and in Hawaii and New Zealand, only around 200 years.

As of 2013, the following are considered Earth's most active volcanoes:

- Kîlauea, the famous Hawaiin volcano, has been in continuous, effusive eruption for thirty years, and has the longest-observedlava lake
- Mount Etna and nearby Stromboli, two Mediterranean volcanoes in "almost continuous eruption" since antiquity
- Mount Yasur, in Vanuatu, has been erupting "nearly continuously" for over 800 years

The longest currently ongoing (but not necessarily continuous) volcanic eruptive phases are:

- Mount Yasur, 111 years
- Mount Etna, 109 years
- Stromboli, 108 years
- Santa María, 101 years
- Sangay, 94 years

Other very active volcanoes include:

- Mount Nyiragongo and its neighbour, Nyamuragira, are Africa's most active volcanoes
- Piton de la Fournaise, in Réunion, erupts frequently enough to be a tourist attraction
- Erta Ale, in the Afar Triangle, has maintained a lava lake since at least 1906
- Mount Erebus, in Antarctica, has maintained a lava lake since at least 1972
- Mount Merapi
- Ol Doinyo Lengai
- Ambrym
- Arenal Volcano
- Pacaya
- Klyuchevskaya Sopka
- Sheveluch

Extinct

Figure: *Fourpeaked volcano, Alaska, in September 2007, after being thought extinct for over 10,000 years.*

Extinct volcanoes are those that scientists consider unlikely to erupt again, because the volcano no longer has a magma supply. Examples of extinct volcanoes are many volcanoes on the Hawaiian – Emperor seamount chain in the Pacific Ocean, Hohentwiel, Shiprock and the Zuidwal volcano in the Netherlands. Edinburgh Castle in Scotland is famously located atop an extinct volcano. Otherwise, whether a volcano is truly extinct is often difficult to determine. Since "supervolcano" calderas can have eruptive lifespans sometimes measured in millions of years, a caldera that has not produced an eruption in tens of thousands of years is likely to be considered dormant instead of extinct. Some volcanologists refer to extinct volcanoes as inactive, though the term is now more commonly used for dormant volcanoes once thought to be extinct.

Dormant

It is difficult to distinguish an extinct volcano from a dormant (inactive) one. Volcanoes are often considered to be extinct if there are no written records of its activity. Nevertheless, volcanoes may remain dormant for a long period of time. For example, Yellowstone has a repose/recharge period of around 700,000 years, and Toba of around 380,000 years. Vesuvius was described by Roman writers as having been covered with gardens and vineyards before its famous eruption of AD 79, which destroyed the towns of Herculaneum and Pompeii. Before its catastrophic eruption of 1991, Pinatubo was an inconspicuous volcano, unknown to most people in the surrounding areas. Two other examples are the long-dormant Soufrière Hills volcano on the island of Montserrat, thought to be extinct before activity resumed in 1995 and Fourpeaked Mountain in Alaska, which, before its September 2006 eruption, had not erupted since before 8000 BC and had long been thought to be extinct.

Technical Classification of Volcanoes

Volcanic-Alert Level: The three common popular classifications of volcanoes can be subjective and some volcanoes thought to have been extinct have erupted again. To help prevent people from falsely believing they are not at risk when living on or near a volcano, countries have adopted new classifications to describe the various levels and stages of volcanic activity. Some alert systems use different numbers or colours to designate the different stages. Other systems use colours and words. Some systems use a combination of both.

Volcano Warning Schemes of the United States

The United States Geological Survey (USGS) has adopted a common system nationwide for characterizing the level of unrest and eruptive activity at volcanoes. The new volcano alert-level system classifies volcanoes now as being in a normal, advisory, watch or warning stage. Additionally, colours are used to denote the amount of ash produced. Details of the US system can be found at Volcano warning schemes of the United States.

Decade Volcanoes

Figure: *Koryaksky volcano towering over Petropavlovsk-Kamchatsky on Kamchatka Peninsula, Far Eastern Russia.*

The Decade Volcanoes are 16 volcanoes identified by the International Association of Volcanology and Chemistry of the Earth's Interior (IAVCEI) as being worthy of particular study in light of their history of large, destructive eruptions and proximity to populated areas. They are named Decade Volcanoes because the project was initiated as part of the United Nations-sponsored International Decade for Natural Disaster Reduction. The 16 current Decade Volcanoes are

- Avachinsky-Koryaksky, Kamchatka, Russia
- Nevado de Colima, Jalisco and Colima, Mexico
- Mount Etna, Sicily, Italy
- Galeras, Nariño, Colombia
- Mauna Loa, Hawaii, USA
- Mount Merapi, Central Java, Indonesia
- Mount Nyiragongo, Democratic Republic of the Congo
- Mount Rainier, Washington, USA
- Sakurajima, Kagoshima Prefecture, Japan
- Santa Maria/Santiaguito, Guatemala

- Santorini, Cyclades, Greece
- Taal Volcano, Luzon, Philippines
- Teide, Canary Islands, Spain
- Ulawun, New Britain, Papua New Guinea
- Mount Unzen, Nagasaki Prefecture, Japan
- Vesuvius, Naples, Italy

Effects of Volcanoes

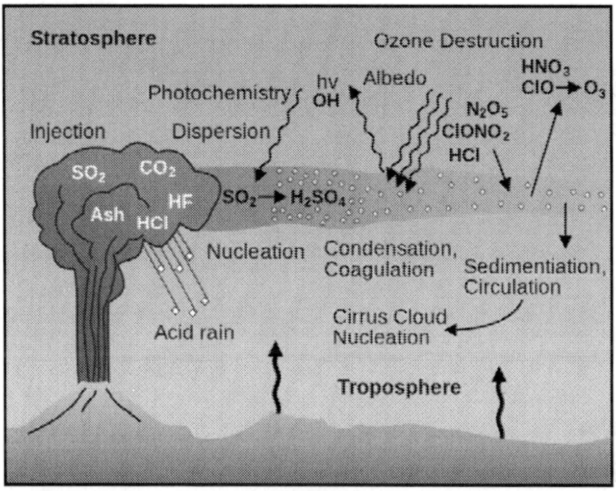

Figure: *Schematic of volcano injection of aerosols and gases.*

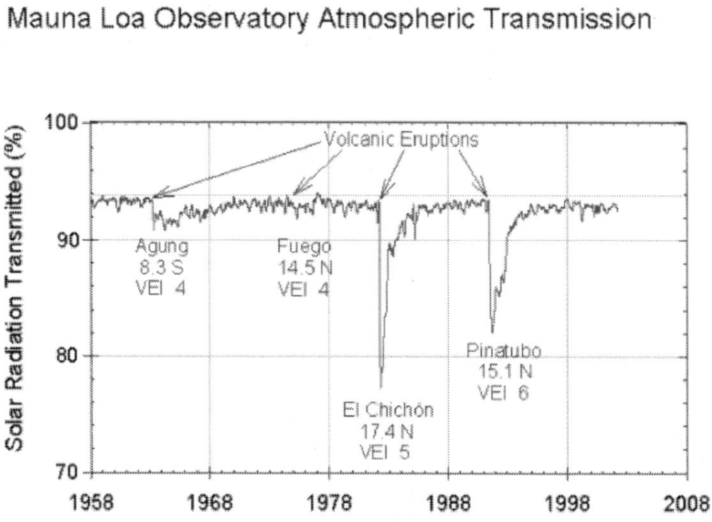

Figure: *Solar radiation graph 1958–2008, showing how the radiation is reduced after major volcanic eruptions.*

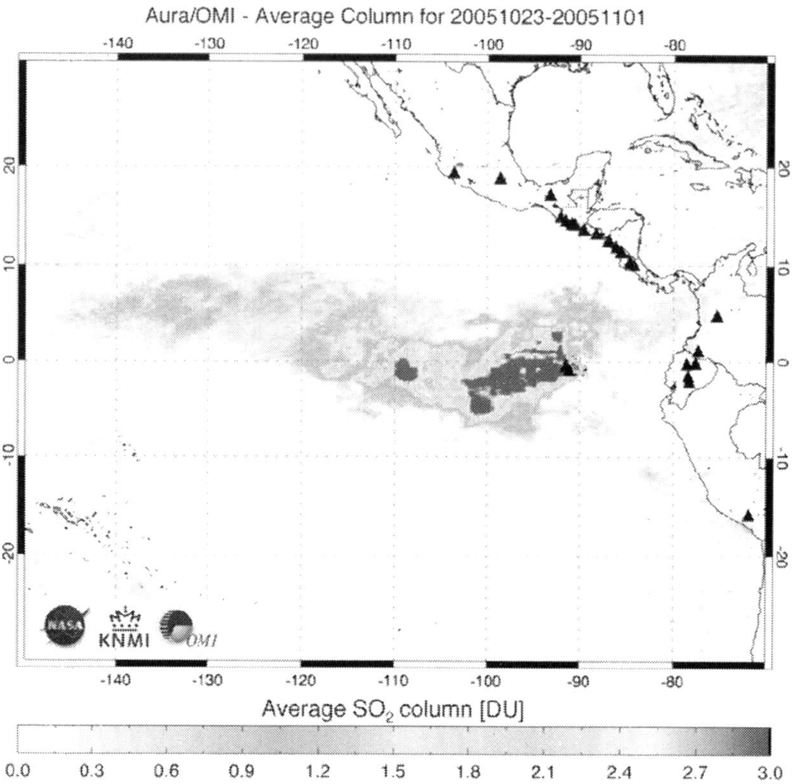

Figure: *Sulfur dioxide concentration over the Sierra Negra Volcano, Galapagos Islands during an eruption in October 2005*

There are many different types of volcanic eruptions and associated activity: phreatic eruptions (steam-generated eruptions), explosive eruption of high-silica lava (e.g., rhyolite), effusive eruption of low-silica lava (e.g., basalt), pyroclastic flows, lahars (debris flow) and carbon dioxide emission. All of these activities can pose a hazard to humans. Earthquakes, hot springs, fumaroles, mud pots and geysers often accompany volcanic activity.

The concentrations of different volcanic gases can vary considerably from one volcano to the next. Water vapor is typically the most abundant volcanic gas, followed by carbon dioxide and sulfur dioxide. Other principal volcanic gases include hydrogen sulfide, hydrogen chloride, and hydrogen fluoride.

A large number of minor and trace gases are also found in volcanic emissions, for example hydrogen, carbon monoxide, halocarbons, organic compounds, and volatile metal chlorides.

Large, explosive volcanic eruptions inject water vapor (H_2O), carbon dioxide (CO_2), sulfur dioxide (SO_2), hydrogen chloride (HCl), hydrogen fluoride (HF) and ash (pulverized rock and pumice) into the stratosphere to heights of 16–32 kilometres (10–20 mi) above the Earth's surface. The most significant impacts from these injections come from the conversion of sulfur dioxide to sulfuric acid (H_2SO_4), which condenses rapidly in the stratosphere to form fine sulfateaerosols. The aerosols increase the Earth's albedo—its reflection of radiation from the Sun back into space – and thus cool the Earth's lower atmosphere or troposphere; however, they also absorb heat radiated up from the Earth, thereby warming the stratosphere. Several eruptions during the past century have caused a decline in the average temperature at the Earth's surface of up to half a degree (Fahrenheit scale) for periods of one to three years – sulfur dioxide from the eruption of Huaynaputina probably caused the Russian famine of 1601–1603.

One proposed volcanic winter happened c. 70,000 years ago following the supereruption of Lake Toba on Sumatra island in Indonesia. According to the Toba catastrophe theory to which some anthropologists and archeologists subscribe, it had global consequences, killing most humans then alive and creating a population bottleneck that affected the genetic inheritance of all humans today. The 1815 eruption of Mount Tambora created global climate anomalies that became known as the "Year Without a Summer" because of the effect on North American and European weather. Agricultural crops failed and livestock died in much of the Northern Hemisphere, resulting in one of the worst famines of the 19th century. The freezing winter of 1740–41, which led to widespread famine in northern Europe, may also owe its origins to a volcanic eruption.

It has been suggested that volcanic activity caused or contributed to the End-Ordovician, Permian-Triassic, Late Devonianmass extinctions, and possibly others. The massive eruptive event which formed the Siberian Traps, one of the largest known volcanic events of the last 500 million years of Earth's geological history, continued for a million years and is considered to be the likely cause of the "Great Dying" about 250 million years ago, which is estimated to have killed 90% of species existing at the time.

The sulfate aerosols also promote complex chemical reactions on their surfaces that alter chlorine and nitrogen chemical species in the stratosphere. This effect, together with increased stratospheric chlorine levels from chlorofluorocarbon pollution, generates chlorine monoxide (ClO), which destroys ozone (O_3). As the aerosols grow and coagulate, they settle down into the upper troposphere where they serve as nuclei for cirrus clouds and further modify the Earth's radiation balance. Most of the hydrogen chloride (HCl) and hydrogen fluoride (HF) are dissolved in water droplets in the eruption cloud and quickly fall to the ground as acid rain. The injected ash also falls rapidly from the stratosphere; most of it is removed within several days to a few weeks. Finally, explosive volcanic eruptions release the greenhouse gas carbon dioxide and thus provide a deep source of carbon for biogeochemical cycles.

Gas emissions from volcanoes are a natural contributor to acid rain. Volcanic activity releases about 130 to 230 teragrams (145 million to 255 million short tons) of carbon dioxide each year. Volcanic eruptions may inject aerosols into the Earth's atmosphere. Large injections may cause visual effects such as unusually colourful sunsets and affect global climate mainly by cooling it. Volcanic eruptions also provide the benefit of adding nutrients to soil through the weathering process of volcanic rocks. These fertile soils assist the growth of plants and various crops. Volcanic eruptions can also create new islands, as the magma cools and solidifies upon contact with the water.

Ash thrown into the air by eruptions can present a hazard to aircraft, especially jet aircraft where the particles can be melted by the high operating temperature; the melted particles then adhere to the turbine blades and alter their shape, disrupting the operation of the turbine. Dangerous encounters in 1982 after the eruption of Galunggung in Indonesia, and 1989 after the eruption of Mount Redoubt in Alaska raised awareness of this phenomenon. Nine Volcanic Ash Advisory Centres were established by the International Civil Aviation Organization to monitor ash clouds and advise pilots accordingly. The 2010 eruptions of Eyjafjallajökull caused major disruptions to air travel in Europe.

Volcanoes on Other Planetary Bodies

The Earth's Moon has no large volcanoes and no current volcanic activity, although recent evidence suggests it may still possess a partially molten core. However, the Moon does have many volcanic features such as maria (the darker patches seen on the moon), rilles and domes.

The planet Venus has a surface that is 90% basalt, indicating that volcanism played a major role in shaping its surface. The planet may have had a major global resurfacing event about 500 million years ago, from what scientists can tell from the density of impact craters on the surface. Lava flows are widespread and forms of volcanism not present on Earth occur as well.

Figure: *The Tvashtar volcano erupts a plume 330 km (205 mi) above the surface of Jupiter's moon Io.*

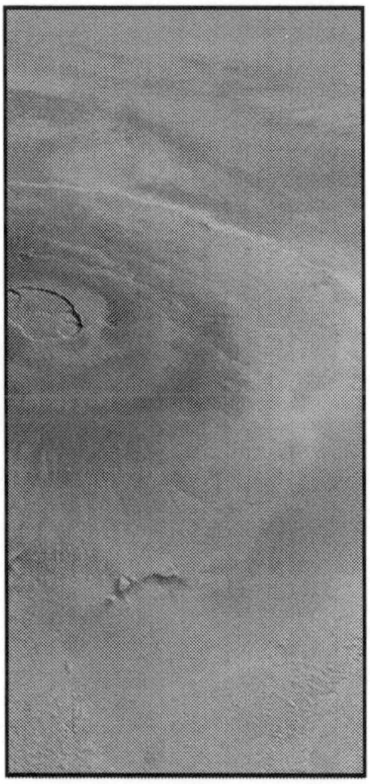

Figure: *Olympus Mons (Latin, "Mount Olympus") is the tallest known mountain in our solar system, located on the planet Mars.*

Changes in the planet's atmosphere and observations of lightning have been attributed to ongoing volcanic eruptions, although there is no confirmation of whether or not Venus is still volcanically active. However, radar sounding by the Magellan probe revealed evidence for comparatively recent volcanic activity at Venus's highest volcano Maat Mons, in the form of ash flows near the summit and on the northern flank.

There are several extinct volcanoes on Mars, four of which are vast shield volcanoes far bigger than any on Earth. They include Arsia Mons, Ascraeus Mons, Hecates Tholus, Olympus Mons, and Pavonis Mons. These volcanoes have been extinct for many millions of years, but the European *Mars Express* spacecraft has found evidence that volcanic activity may have occurred on Mars in the recent past as well.

Jupiter's moonIo is the most volcanically active object in the solar system because of tidal interaction with Jupiter. It is covered with volcanoes that erupt sulfur, sulfur dioxide and silicate rock, and as a result, Io is constantly being resurfaced. Its lavas are the hottest known anywhere in the solar system, with temperatures exceeding 1,800 K (1,500 °C). In February 2001, the largest recorded volcanic eruptions in the solar system occurred on Io. Europa, the smallest of Jupiter's Galilean moons, also appears to have an active volcanic system, except that its volcanic activity is entirely in the form of water, which freezes into ice on the frigid surface. This process is known as cryovolcanism, and is apparently most common on the moons of the outer planets of the solar system.

In 1989 the Voyager 2 spacecraft observed cryovolcanoes (ice volcanoes) on Triton, a moon of Neptune, and in 2005 the Cassini–Huygens probe photographed fountains of frozen particles erupting from Enceladus, a moon of Saturn. The ejecta may be composed of water, liquid nitrogen, dust, or methane compounds. Cassini–Huygens also found evidence of a methane-spewing cryovolcano on the Saturnian moon Titan, which is believed to be a significant source of the methane found in its atmosphere. It is theorized that cryovolcanism may also be present on the Kuiper Belt Object Quaoar.

A 2010 study of the exoplanet COROT-7b, which was detected by transit in 2009, studied that tidal heating from the host star very close to the planet and neighbouring planets could generate intense volcanic activity similar to Io.

Traditional Beliefs about Volcanoes

Many ancient accounts ascribe volcanic eruptions to supernatural causes, such as the actions of gods or demigods. To the ancient Greeks, volcanoes' capricious power could only be explained as acts of the gods, while 16th/17th-century German astronomer Johannes Kepler believed they were ducts for the Earth's tears. One early idea counter to this was proposed by Jesuit Athanasius Kircher (1602–1680), who witnessed eruptions of Mount Etna and Stromboli, then visited the crater of Vesuvius and published his view of an Earth with a central fire connected to numerous others caused by the burning of sulfur, bitumen and coal.

Various explanations were proposed for volcano behaviour before the modern understanding of the Earth's mantle structure as a semisolid material was developed. For decades after awareness that compression and radioactive materials may be heat sources, their contributions were specifically discounted. Volcanic action was often attributed to chemical reactions and a thin layer of molten rock near the surface.

7

Plate Tectonics

Plate tectonics (from the Late Latin *tectonicus*, from the Greek: ôåêôïíéêüò "pertaining to building") is a scientific theory that describes the large-scale motions of Earth's lithosphere. The model builds on the concepts of continental drift, developed during the first few decades of the 20th century. The geoscientific community accepted the theory after the concepts of seafloor spreading were developed in the late 1950s and early 1960s.

The lithosphere is broken up into tectonic plates. On Earth, there are seven or eight major plates (depending on how they are defined) and many minor plates. Where plates meet, their relative motion determines the type of boundary: convergent, divergent, or transform. Earthquakes, volcanic activity, mountain-building, and oceanic trench formation occur along these plate boundaries. The lateral relative movement of the plates typically varies from zero to 100 mm annually.

Tectonic plates are composed of oceanic lithosphere and thicker continental lithosphere, each topped by its own kind of crust. Along convergent boundaries, subduction carries plates into the mantle; the material lost is roughly balanced by the formation of new (oceanic) crust along divergent margins by seafloor spreading. In this way, the total surface of the globe remains the same. This prediction of plate tectonics is also referred to as the conveyor belt principle. Earlier theories (that still have some supporters) proposed gradual shrinking (contraction) or gradual expansion of the globe.

Tectonic plates are able to move because the Earth's lithosphere has a higher strength than the underlying asthenosphere. Lateral density variations in the mantle result in convection. Plate movement is thought to be driven by a combination of the motion of the seafloor away from the spreading ridge (due to variations in topography and density of the crust, which result in differences in gravitational forces) and drag, downward suction, at the subduction zones. Another explanation lies in the different forces generated by the rotation of the globe and the tidal forces of the Sun and the Moon. The relative importance of each of these factors is unclear, and is still subject to debate.

Key Principles

The outer layers of the Earth are divided into lithosphere and asthenosphere. This is based on differences in mechanical properties and in the method for the transfer of heat. Mechanically, the lithosphere is cooler and more rigid, while the asthenosphere is hotter and flows more easily. In terms of heat transfer, the lithosphere loses heat by conduction, whereas the asthenosphere also transfers heat by convection and has a nearly adiabatic temperature gradient. This division should not be confused with the *chemical* subdivision of these same layers into the mantle (comprising both the asthenosphere and the mantle portion of the lithosphere) and the crust: a given piece of mantle may be part of the lithosphere or the asthenosphere at different times, depending on its temperature and pressure.

The key principle of plate tectonics is that the lithosphere exists as separate and distinct *tectonic plates*, which ride on the fluid-like (visco-elastic solid) asthenosphere. Plate motions range up to a typical 10–40 mm/a (Mid-Atlantic Ridge; about as fast as fingernails grow), to about 160 mm/a (Nazca Plate; about as fast as hair grows).

Tectonic lithosphere plates consist of lithospheric mantle overlain by either or both of two types of crustal material: oceanic crust (in older texts called *sima* from silicon and magnesium) and continental crust (*sial* from silicon and aluminium). Average oceanic lithosphere is typically 100 km (62 mi) thick; its thickness is a function of its age: as time passes, it conductively cools and subjacent cooling mantle is added to its base. Because it is formed at mid-ocean ridges and spreads outwards, its thickness is therefore a function of its distance from the mid-ocean ridge where it was formed. For a typical distance oceanic lithosphere must travel before being subducted, the thickness varies from about 6 km (4 mi) thick at mid-ocean ridges to greater than 100 km (62 mi) at subduction zones; for shorter or longer distances, the subduction zone (and therefore also the mean) thickness becomes smaller or larger, respectively. Continental lithosphere is typically

~200 km thick, though this also varies considerably between basins, mountain ranges, and stable cratonic interiors of continents. The two types of crust also differ in thickness, with continental crust being considerably thicker than oceanic (35 km vs. 6 km).

The location where two plates meet is called a *plate boundary*, and plate boundaries are commonly associated with geological events such as earthquakes and the creation of topographic features such as mountains, volcanoes, mid-ocean ridges, and oceanic trenches. The majority of the world's active volcanoes occur along plate boundaries, with the Pacific Plate's Ring of Fire being most active and most widely known. These boundaries are discussed in further detail below. Some volcanoes occur in the interiors of plates, and these have been variously attributed to internal plate deformation and to mantle plumes.

As explained above, tectonic plates can include continental crust or oceanic crust, and many plates contain both. For example, the African Plate includes the continent and parts of the floor of the Atlantic and Indian Oceans. The distinction between oceanic crust and continental crust is based on their modes of formation. Oceanic crust is formed at sea-floor spreading centres, and continental crust is formed through arc volcanism and accretion of terranes through tectonic processes; though some of these terranes may contain ophiolite sequences, which are pieces of oceanic crust, these are considered part of the continent when they exit the standard cycle of formation and spreading centres and subduction beneath continents. Oceanic crust is also denser than continental crust owing to their different compositions. Oceanic crust is denser because it has less silicon and more heavier elements ("mafic") than continental crust ("felsic"). As a result of this density stratification, oceanic crust generally lies below sea level (for example most of the Pacific Plate), while the continental crust buoyantly projects above sea level.

Types of Plate Boundaries

Three types of plate boundaries exist, with a fourth, mixed type, characterized by the way the plates move relative to each other. They are associated with different types of surface phenomena. The different types of plate boundaries are:

1. *Transform boundaries (Conservative)* occur where plates slide or, perhaps more accurately, grind past each other along transform faults. The relative motion of the two plates is either sinistral (left side towards the observer) or dextral (right side towards the observer). The San Andreas Fault in California

is an example of a transform boundary exhibiting dextral motion.

2. *Divergent boundaries (Constructive)* occur where two plates slide apart from each other. Mid-ocean ridges (e.g., Mid-Atlantic Ridge) and active zones of rifting (such as Africa's East African Rift) are both examples of divergent boundaries.
3. *Convergent boundaries (Destructive)* (or *active margins*) occur where two plates slide towards each other commonly forming either a subduction zone (if one plate moves underneath the other) or a continental collision (if the two plates contain continental crust). Deep marine trenches are typically associated with subduction zones, and the basins that develop along the active boundary are often called "foreland basins". The subducting slab contains many hydrous minerals, which release their water on heating; this water then causes the mantle to melt, producing volcanism. Examples of this are the Andes mountain range in South America and the Japanese island arc.
4. *Plate boundary zones* occur where the effects of the interactions are unclear and the boundaries, usually occurring along a broad belt, are not well defined, and may show various types of movements in different episodes.

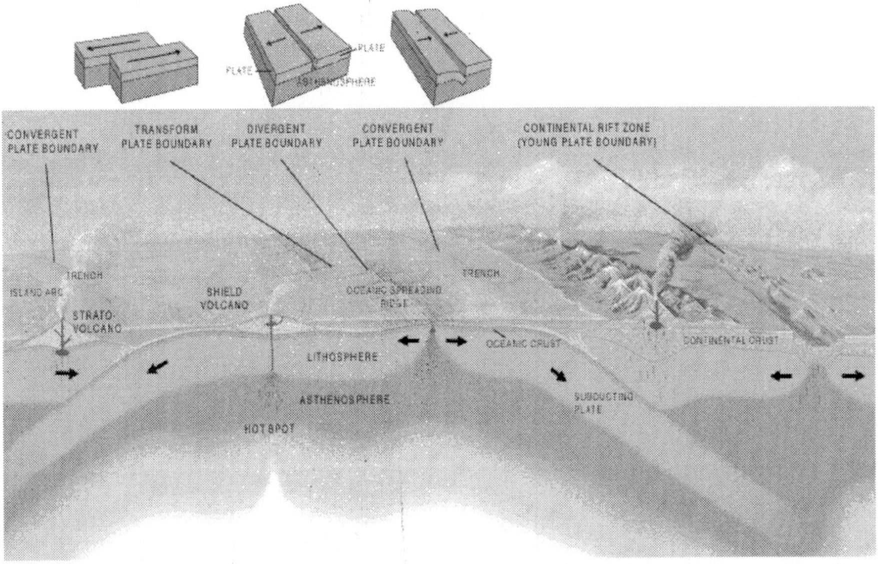

Figure: *Three types of plate boundary.*

Driving Forces of Plate Motion

Plate tectonics is basically a kinematic phenomenon: Earth scientists agree on the observation and deduction that the plates have moved with respect to one another, and debate and find agreements as to how and when. But still, a major question remains as to what the motor behind this movement is - the geodynamic mechanism - and here science diverges in different theories.

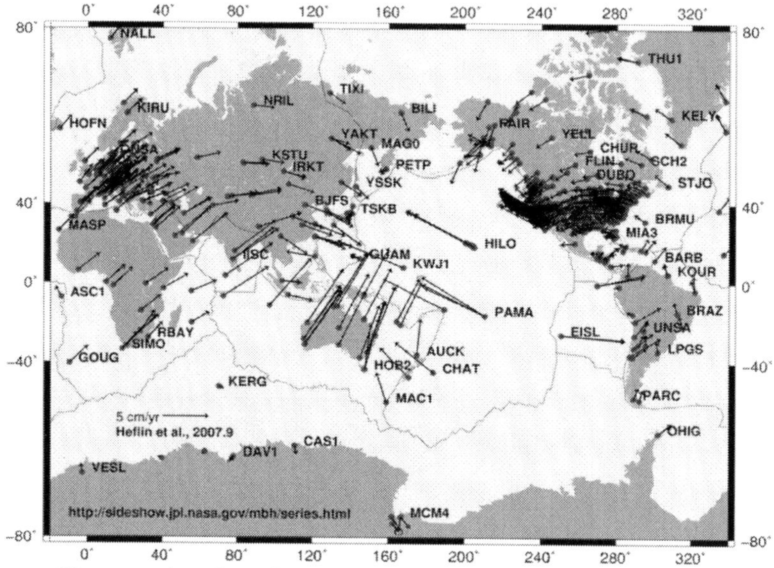

Figure: *Plate motion based on Global Positioning System (GPS) satellite data from NASA JPL. The vectors show direction and magnitude of motion.*

Generally, it is accepted that tectonic plates are able to move because of the relative density of oceanic lithosphere and the relative weakness of the asthenosphere. Dissipation of heat from the mantle is acknowledged to be the original source of energy driving plate tectonics, through convection or large scale upwelling and doming. As a consequence, in the current view, although it is still a matter of some debate, because of the excess density of the oceanic lithosphere sinking in subduction zones a powerful source of plate motion is generated. When the new crust forms at mid-ocean ridges, this oceanic lithosphere is initially less dense than the underlying asthenosphere, but it becomes denser with age, as it conductively cools and thickens. The greater density of old lithosphere relative to the underlying asthenosphere allows it to sink into the deep mantle at subduction zones, providing most of the driving force for plate motions. The weakness of the asthenosphere allows the tectonic plates to move easily towards a subduction zone. Although subduction is believed to be the strongest force driving plate motions, it cannot be the only force since there are plates such as the North American Plate which are moving, yet are nowhere being subducted. The same is true for the enormous Eurasian Plate. The sources of plate motion are a matter of intensive research and discussion among Earth scientists. One of the main points is that the kinematic pattern of the movements itself should be separated clearly from the possible geodynamic mechanism that is invoked as the driving force of the observed movements, as some patterns may be explained by more than one mechanism. Basically, the driving forces that are advocated at the moment, can be divided in three categories: mantle dynamics related, gravity related (mostly secondary forces), and Earth rotation related.

Driving Forces Related to Mantle Dynamics

For a considerable period of around 25 years (last quarter of the twentieth century) the leading theory envisaged large scale convection currents in the upper mantle which are transmitted through the asthenosphere as the main driving force of the tectonic plates. This theory was launched by Arthur Holmes and some forerunners in the 1930s and was immediately recognized as the solution for the acceptance of the theory discussed since its occurrence in the papers of Alfred Wegener in the early years of the century. It was, though, long debated because the leading ("fixist") theory was still envisaging a static Earth without moving continents, up until the major break–throughs in the early sixties.

Two– and three–dimensional imaging of the Earth's interior (seismic tomography) shows that there is a laterally varying density distribution throughout the mantle. Such density variations can be material (from rock chemistry), mineral (from variations in mineral structures), or thermal (through thermal expansion and contraction from heat energy). The manifestation of this varying lateral density is mantle convection from buoyancy forces.

How mantle convection relates directly and indirectly to the motion of the plates is a matter of ongoing study and discussion in geodynamics. Somehow, this energy must be transferred to the lithosphere for tectonic plates to move. There are essentially two types of forces that are thought to influence plate motion: friction and gravity.

- *Basal drag (friction):* The plate motion is in this way driven by friction between the convection currents in the asthenosphere and the more rigid overlying floating lithosphere.
- *Slab suction (gravity):* Local convection currents exert a downward frictional pull on plates in subduction zones at ocean trenches. Slab suction may occur in a geodynamic setting wherein basal tractions continue to act on the plate as it dives into the mantle (although perhaps to a greater extent acting on both the under and upper side of the slab).

Lately, the convection theory is much debated as modern techniques based on 3D seismic tomography of imaging the internal structure of the Earth's mantle still fail to recognize these predicted large scale convection cells. Therefore, alternative views have been proposed:

In the theory of plume tectonics developed during the 1990s, a modified concept of mantle convection currents is used, related to super plumes rising from the deeper mantle which would be the drivers or the substitutes of the major convection cells. These ideas, which find their roots in the early 1930s with the so-called "fixistic" ideas of the European and Russian Earth Science Schools, find resonance in the modern theories which envisage hot spots/mantle plumes in the mantle which remain fixed and are overridden by oceanic and continental lithosphere plates during time, and leave their traces in the geological record (though these phenomena are not invoked as real driving mechanisms, but rather as a modulator). The modern theories that continue building on the older mantle doming concepts and see the movements of the plates a secondary phenomena, are beyond the scope of this page and are discussed elsewhere for example on the plume tectonics page.

Another suggestion is that the mantle flows neither in cells nor large plumes, but rather as a series of channels just below the Earth's crust, which then provide basal friction to the lithosphere. This theory is called "surge tectonics" and became quite popular in geophysics and geodynamics during the 1980s and 1990s.

Driving Forces Related to Gravity

Gravity related forces are usually invoked as secondary phenomena within the framework of a more general driving mechanism such as the various forms of mantle dynamics described above.

Gravitational sliding away from a spreading ridge: According to many authors, plate motion is driven by the higher elevation of plates at ocean ridges. As oceanic lithosphere is formed at spreading ridges from hot mantle material, it gradually cools and thickens with age (and thus distance from the ridge). Cool oceanic lithosphere is significantly denser than the hot mantle material from which it is derived and so with increasing thickness it gradually subsides into the mantle to compensate the greater load. The result is a slight lateral incline with distance from the ridge axis.

This force is regarded as a secondary force and is often referred to as "ridge push". This is a misnomer as nothing is "pushing" horizontally and tensional features are dominant along ridges. It is more accurate to refer to this mechanism as gravitational sliding as variable topography across the totality of the plate can vary considerably and the topography of spreading ridges is only the most prominent feature. Other mechanisms generating this gravitational secondary force include flexural bulging of the lithosphere before it dives underneath an adjacent plate, which produces a clear topographical feature that can offset or at least affect the influence of topographical ocean ridges, and mantle plumes and hot spots, which are postulated to impinge on the underside of tectonic plates.

Slab-pull: Current scientific opinion is that the asthenosphere is insufficiently competent or rigid to directly cause motion by friction along the base of the lithosphere. Slab pull is therefore most widely thought to be the greatest force acting on the plates. In this current understanding, plate motion is mostly driven by the weight of cold, dense plates sinking into the mantle at trenches. Recent models indicate that trench suction plays an important role as well. However, as the North American Plate is nowhere being subducted, yet it is in motion presents a problem. The same holds for the African, Eurasian, and Antarctic plates.

Gravitational sliding away from mantle doming: According to older theories one of the driving mechanisms of the plates is the existence of large scale asthenosphere/mantle domes, which cause the gravitational sliding of lithosphere plates away from them. This gravitational sliding represents a secondary phenomenon of this, basically vertically oriented mechanism. This can act on various scales, from the small scale of one island arc up to the larger scale of an entire ocean basin.

Driving Forces Related to Earth Rotation

Alfred Wegener, being a meteorologist, had proposed tidal forces and pole flight force as main driving mechanisms for continental drift. However, these forces were considered far too small to cause continental motion as the concept then was of continents plowing through oceanic crust. Therefore, Wegener converted to convection currents as the main driving force in the last edition of his book in 1929.

In the plate tectonics context (accepted since the seafloor spreading proposals of Heezen, Hess, Dietz, Morley, Vine and Matthews during the early 1960s) though, oceanic crust is in motion *with* the continents which caused the proposals related to Earth rotation to be reconsidered. In more recent literature, these driving forces are:

1. Tidal drag due to the gravitational force the Moon (and the Sun) exerts on the crust of the Earth
2. Shear strain of the Earth globe due to N-S compression related to the rotation and modulations of it;
3. Pole flight force: equatorial drift due to rotation and centrifugal effects: tendency of the plates to move from the poles to the equator ("*Polflucht*");
4. Coriolis effect acting on plates when they move around the globe;
5. Global deformation of the geoid due to small displacements of rotational pole with respect to the Earth crust;
6. Other smaller deformation effects of the crust due to wobbles and spin movements of the Earth rotation on a smaller time scale.

For these mechanisms to be overall valid, systematic relationships should exist all over the globe between the orientation and kinematics of deformation, and the geographical latitudinal and longitudinal grid of the Earth itself. Ironically, these systematic relations studies in the second half of the nineteenth century and the first half of the twentieth century do underline exactly the opposite: that the plates had not moved in time, that the deformation grid was fixed with respect to the Earth equator and axis, and that gravitational driving forces were generally acting vertically and caused only local horizontal movements (the so-called pre-plate tectonic, "fixist theories"). Later studies (discussed below on this page) therefore invoked many of the relationships recognized during this pre-plate tectonics period, to support their theories.

Of the many forces discussed in this paragraph, tidal force is still highly debated and defended as a possible principle driving force, whereas the other forces are used or in global geodynamic models not using the plate tectonics concepts (therefore beyond the discussions treated in this section), or proposed as minor modulations within the overall plate tectonics model.

In 1973, George W. Moore of the USGS and R. C. Bostrom presented evidence for a general westward drift of the Earth's lithosphere with respect to the mantle, and, therefore, tidal forces or tidal lag or "friction" due to the Earth's rotation and the forces acting upon it by the Moon being a driving force for plate tectonics: as the Earth spins eastward beneath the moon, the moon's gravity ever so slightly pulls the Earth's surface layer back westward, just as proposed by Alfred Wegener. In a more recent 2006 study, scientists reviewed and advocated these earlier proposed ideas. It has also been suggested recently in Lovett (2006) that this observation may also explain why Venus and Mars have no plate tectonics, since Venus has no moon and Mars' moons are too small to have significant tidal effects on Mars. In a recent paper, it was suggested that, on the other hand, it can easily be observed that many plates are moving north and eastward, and that the dominantly westward motion of the Pacific ocean basins derives simply from the eastward bias of the Pacific spreading centre (which is not a predicted manifestation of such lunar forces). In the same paper the authors admit, however, that relative to the lower mantle, there is a slight westward component in the motions of all the plates. They demonstrated though that the westward drift, seen only for the past 30 Ma, is attributed to the increased dominance of the steadily growing and accelerating Pacific plate. The debate is still open.

Relative Significance of Each Driving Force Mechanism

The actual vector of a plate's motion must necessarily be a function of all the forces acting on the plate. However, therein remains the problem regarding what degree each process contributes to the motion of each tectonic plate.

The diversity of geodynamic settings and properties of each plate must clearly result in differences in the degree to which such processes are actively driving the plates. One method of dealing with this problem is to consider the relative rate at which each plate is moving and to consider the available evidence of each driving force on the plate as far as possible.

One of the most significant correlations found is that lithospheric plates attached to downgoing (subducting) plates move much faster than plates not attached to subducting plates. The Pacific plate, for instance, is essentially surrounded by zones of subduction (the so-called Ring of Fire) and moves much faster than the plates of the Atlantic basin, which are attached (perhaps one could say 'welded') to adjacent continents instead of subducting plates. It is thus thought that forces associated with the downgoing plate (slab pull and slab suction) are the driving forces which determine the motion of plates, except for those plates which are not being subducted. The driving forces of plate motion continue to be active subjects of on-going research within geophysics and tectonophysics.

Development of the Theory

In line with other previous and contemporaneous proposals, in 1912 the meteorologist Alfred Wegener amply described what he called continental drift, expanded in his 1915 book *The Origin of Continents and Oceans* and the scientific debate started that would end up fifty years later in the theory of plate tectonics. Starting from the idea (also expressed by his forerunners) that the present continents once formed a single land mass (which was called Pangea later on) that drifted apart, thus releasing the continents from the Earth's mantle and likening them to "icebergs" of low density granite floating on a sea of denser basalt. Supporting evidence for the idea came from the dove-tailing outlines of South America's east coast and Africa's west coast, and from the matching of the rock formations along these edges. Confirmation of their previous contiguous nature also came from the fossil plants *Glossopteris* and *Gangamopteris*, and the therapsid or mammal-like reptile *Lystrosaurus*, all widely distributed over South America, Africa, Antarctica, India and Australia. The evidence for such an erstwhile joining of these continents was patent to field geologists working in the southern hemisphere. The South African Alex du Toit put together a mass of such information in his 1937 publication *Our Wandering Continents*, and went further than Wegener in recognising the strong links between the Gondwana fragments.

But without detailed evidence and a force sufficient to drive the movement, the theory was not generally accepted: the Earth might have a solid crust and mantle and a liquid core, but there seemed to be no way that portions of the crust could move around. Distinguished scientists, such as Harold Jeffreys and Charles Schuchert, were outspoken critics of continental drift.

Despite much opposition, the view of continental drift gained support and a lively debate started between "drifters" or "mobilists" (proponents of the theory) and "fixists" (opponents). During the 1920s, 1930s and 1940s, the former reached important milestones proposing that convection currents might have driven the plate movements, and that spreading may have occurred below the sea within the oceanic crust. Concepts close to the elements now incorporated in plate tectonics were proposed by geophysicists and geologists (both fixists and mobilists) like Vening-Meinesz, Holmes, and Umbgrove.

One of the first pieces of geophysical evidence that was used to support the movement of lithospheric plates came from paleomagnetism. This is based on the fact that rocks of different ages show a variable magnetic field direction, evidenced by studies since the mid–nineteenth century. The magnetic north and south poles reverse through time, and, especially important in paleotectonic studies, the relative position of the magnetic north pole varies through time. Initially, during the first half of the twentieth century, the latter phenomenon was explained by introducing what was called "polar wander", i.e., it was assumed that the north pole location had been shifting through time. An alternative explanation, though, was that the continents had moved (shifted and rotated) relative to the north pole, and each continent, in fact, shows its own "polar wander path". During the late 1950s it was successfully shown on two occasions that these data could show the validity of continental drift: by Keith Runcorn in a paper in 1956, and by Warren Carey in a symposium held in March 1956.

The second piece of evidence in support of continental drift came during the late 1950s and early 60s from data on the bathymetry of the deep ocean floors and the nature of the oceanic crust such as magnetic properties and, more generally, with the development of marine geology which gave evidence for the association of seafloor spreading along the mid-oceanic ridges and magnetic field reversals, published between 1959 and 1963 by Heezen, Dietz, Hess, Mason, Vine & Matthews, and Morley.

Simultaneous advances in early seismic imaging techniques in and around Wadati-Benioff zones along the trenches bounding many continental margins, together with many other geophysical (e.g. gravimetric) and geological observations, showed how the oceanic crust could disappear into the mantle, providing the mechanism to balance the extension of the ocean basins with shortening along its margins.

All this evidence, both from the ocean floor and from the continental margins, made it clear around 1965 that continental drift was feasible and the theory of plate tectonics, which was defined in a series of papers between 1965 and 1967, was born, with all its extraordinary explanatory and predictive power. The theory revolutionized the Earth sciences, explaining a diverse range of geological phenomena and their implications in other studies such as paleogeography and paleobiology.

Continental Drift

In the late 19th and early 20th centuries, geologists assumed that the Earth's major features were fixed, and that most geologic features such as basin development and mountain ranges could be explained by vertical crustal movement, described in what is called the geosynclinal theory. Generally, this was placed in the context of a contracting planet Earth due to heat loss in the course of a relatively short geological time.

It was observed as early as 1596 that the opposite coasts of the Atlantic Ocean—or, more precisely, the edges of the continental shelves—have similar shapes and seem to have once fitted together.

Since that time many theories were proposed to explain this apparent complementarity, but the assumption of a solid Earth made these various proposals difficult to accept.

The discovery of radioactivity and its associated heating properties in 1895 prompted a re-examination of the apparent age of the Earth. This had previously been estimated by its cooling rate and assumption the Earth's surface radiated like a black body. Those calculations had implied that, even if it started at red heat, the Earth would have dropped to its present temperature in a few tens of millions of years. Armed with the knowledge of a new heat source, scientists realized that the Earth would be much older, and that its core was still sufficiently hot to be liquid.

By 1915, after having published a first article in 1912, Alfred Wegener was making serious arguments for the idea of continental drift in the first edition of *The Origin of Continents and Oceans*. In that book (re-issued in four successive editions up to the final one in 1936), he noted how the east coast of South America and the west coast of Africa looked as if they were once attached. Wegener was not the first to note this (Abraham Ortelius, Snider-Pellegrini, Eduard Suess, Roberto Mantovani and Frank Bursley Taylor preceded him just to mention a few), but he was the first to marshal significant fossil and paleo-topographical and climatological evidence to support this simple observation (and was supported in this by researchers such as Alex du Toit). Furthermore, when the rock strata of the margins of separate continents are very similar it suggests that these rocks were formed in the same way, implying that they were joined initially. For instance, parts of Scotland and Ireland contain rocks very similar to those found in Newfoundland and New Brunswick. Furthermore, the Caledonian Mountains of Europe and parts of the Appalachian Mountains of North America are very similar in structure and lithology.

However, his ideas were not taken seriously by many geologists, who pointed out that there was no apparent mechanism for continental drift. Specifically, they did not see how continental rock could plow through the much denser rock that makes up oceanic crust. Wegener could not explain the force that drove continental drift, and his vindication did not come until after his death in 1930.

Floating Continents, Paleomagnetism, and Seismicity Zones

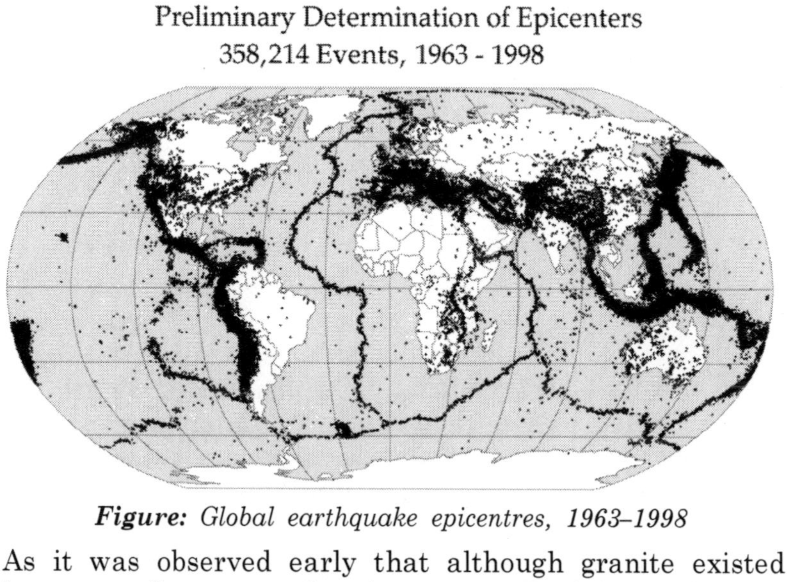

Figure: *Global earthquake epicentres, 1963–1998*

As it was observed early that although granite existed on continents, seafloor seemed to be composed of denser basalt, the prevailing concept during the first half of the twentieth century was that there were two types of crust, named "sial" (continental type crust) and "sima" (oceanic type crust). Furthermore, it was supposed that a static shells of strata was present under the continents. It therefore looked apparent that a layer of basalt (sial) underlies the continental rocks.

However, based on abnormalities in plumb line deflection by the Andes in Peru, Pierre Bouguer had deduced that less-dense mountains must have a downward projection into the denser layer underneath. The concept that mountains had "roots" was confirmed by George B. Airy a hundred years later, during study of Himalayan gravitation, and seismic studies detected corresponding density variations. Therefore, by the mid–1950s, the question remained unresolved as to whether mountain roots were clenched in surrounding basalt or were floating on it like an iceberg.

During the 20th century, improvements in and greater use of seismic instruments such as seismographs enabled scientists to learn that earthquakes tend to be concentrated in specific areas, most notably along the oceanic trenches and spreading ridges. By the late 1920s, seismologists were beginning to identify several prominent earthquake zones parallel to the trenches that typically were inclined 40–60° from the horizontal and extended several hundred kilometres into the Earth. These zones later became known as Wadati-Benioff zones, or simply Benioff zones, in honor of the seismologists who first recognized them, Kiyoo Wadati of Japan and Hugo Benioff of the United States. The study of global seismicity greatly advanced in the 1960s with the establishment of the Worldwide Standardized Seismograph Network (WWSSN) to monitor the compliance of the 1963 treaty banning above-ground testing of nuclear weapons. The much improved data from the WWSSN instruments allowed seismologists to map precisely the zones of earthquake concentration world wide.

Meanwhile, debates developed around the phenomena of polar wander. Since the early debates of continental drift, scientists had discussed and used evidence that polar drift had occurred because continents seemed to have moved through different climatic zones during the past. Furthermore, paleomagnetic data had shown that the magnetic pole had also shifted during time. Reasoning in an opposite way, the continents might have shifted and rotated, while the pole remained relatively fixed. The first time the evidence of magnetic polar wander was used to support the movements of continents was in a paper by Keith Runcorn in 1956, and successive papers by him and his students Ted Irving (who was actually the first to be convinced of the fact that paleomagnetism supported continental drift) and Ken Creer.

This was immediately followed by a symposium in Tasmania in March 1956. In this symposium, the evidence was used in the theory of an expansion of the global crust. In this hypothesis the shifting of the continents can be simply explained by a large increase in size of the Earth since its formation. However, this was unsatisfactory because its supporters could offer no convincing mechanism to produce a significant expansion of the Earth. Certainly there is no evidence that the moon has expanded in the past 3 billion years; other work would soon show that the evidence was equally in support of continental drift on a globe with a stable radius.

During the thirties up to the late fifties, works by Vening-Meinesz, Holmes, Umbgrove, and numerous others outlined concepts that were close or nearly identical to modern plate tectonics theory. In particular, the English geologist Arthur Holmes proposed in 1920 that plate junctions might lie beneath the sea, and in 1928 that convection currents within the mantle might be the driving force. Often, these contributions are forgotten because:

- At the time, continental drift was not accepted.
- Some of these ideas were discussed in the context of abandoned fixistic ideas of a deforming globe without continental drift or an expanding Earth.
- They were published during an episode of extreme political and economic instability that hampered scientific communication.
- Many were published by European scientists and at first not mentioned or given little credit in the papers on sea floor spreading published by the American researchers in the 1960s.

Mid-Oceanic Ridge Spreading and Convection

In 1947, a team of scientists led by Maurice Ewing utilizing the Woods Hole Oceanographic Institution's research vessel *Atlantis* and an array of instruments, confirmed the existence of a rise in the central Atlantic Ocean, and found that the floor of the seabed beneath the layer of sediments consisted of basalt, not the granite which is the main constituent of continents. They also found that the oceanic crust was much thinner than continental crust. All these new findings raised important and intriguing questions.

The new data that had been collected on the ocean basins also showed particular characteristics regarding the bathymetry. One of the major outcomes of these datasets was that all along the globe, a system of mid-oceanic ridges was detected. An important conclusion was that along this system, new ocean floor was being created, which led to the concept of the "Great Global Rift". This was described in the crucial paper of Bruce Heezen (1960), which would trigger a real revolution in thinking. A profound consequence of seafloor spreading is that new crust was, and still is, being continually created along the oceanic ridges. Therefore, Heezen advocated the so-called "expanding Earth" hypothesis of S. Warren Carey. So, still the question remained: how can new crust be continuously added along the oceanic ridges without increasing the size of the Earth? In reality, this question had been solved already by numerous scientists during the forties and the fifties, like Arthur Holmes, Vening-Meinesz, Coates and many others: The crust in excess disappeared along what were called the oceanic trenches, where so-called "subduction" occurred. Therefore, when various scientists during the early sixties started to reason on the data at their disposal regarding the ocean floor, the pieces of the theory quickly fell into place.

The question particularly intrigued Harry Hammond Hess, a Princeton University geologist and a Naval Reserve Rear Admiral, and Robert S. Dietz, a scientist with the U.S. Coast and Geodetic Survey who first coined the term *seafloor spreading*. Dietz and Hess (the former published the same idea one year earlier in *Nature*, but priority belongs to Hess who had already distributed an unpublished manuscript of his 1962 article by 1960) were among the small handful who really understood the broad implications of sea floor spreading and how it would eventually agree with the, at that time, unconventional and unaccepted ideas of continental drift and the elegant and mobilistic models proposed by previous workers like Holmes.

In the same year, Robert R. Coats of the U.S. Geological Survey described the main features of island arc subduction in the Aleutian Islands. His paper, though little–noted (and even ridiculed) at the time, has since been called "seminal" and "prescient". In reality, it actually shows that the work by the European scientists on island arcs and mountain belts performed and published during the 1930s up until the 1950s was applied and appreciated also in the United States.

If the Earth's crust was expanding along the oceanic ridges, Hess and Dietz reasoned like Holmes and others before them, it must be shrinking elsewhere. Hess followed Heezen, suggesting that new oceanic crust continuously spreads away from the ridges in a conveyor belt–like motion. And, using the mobilistic concepts developed before, he correctly concluded that many millions of years later, the oceanic crust eventually descends along the continental margins where oceanic trenches – very deep, narrow canyons – are formed, e.g. along the rim of the Pacific Ocean basin. The important step Hess made was that convection currents would be the driving force in this process, arriving at the same conclusions as Holmes had decades before with the only difference that the thinning of the ocean crust was performed using Heezen's mechanism of spreading along the ridges. Hess therefore concluded that the Atlantic Ocean was expanding while the Pacific Ocean was shrinking. As old oceanic crust is "consumed" in the trenches, (like Holmes and others, he thought this was done by thickening of the continental lithosphere, not, as now understood, by underthrusting at a larger scale of the oceanic crust itself into the mantle) new magma rises and erupts along the spreading ridges to form new crust. In effect, the ocean basins are perpetually being "recycled," with the creation of new crust and the destruction of old oceanic lithosphere occurring simultaneously. Thus, the new mobilistic concepts neatly explained why the Earth does not get bigger with sea floor spreading, why there is so little sediment accumulation on the ocean floor, and why oceanic rocks are much younger than continental rocks.

Magnetic Striping

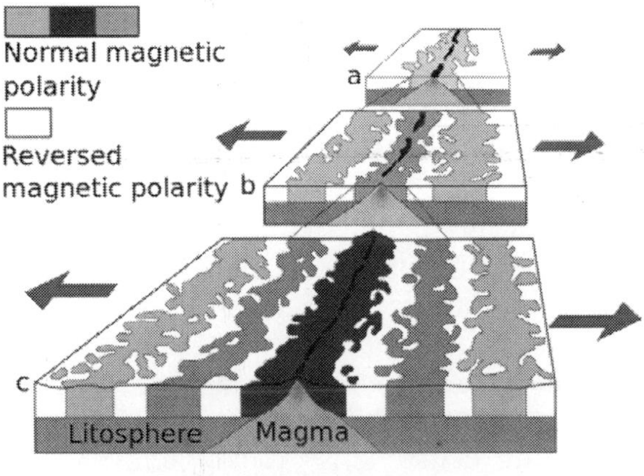

Figure: *Seafloor magnetic striping.*

Figure: *A demonstration of magnetic striping. (The darker the colour is the closer it is to normal polarity)*

Beginning in the 1950s, scientists like Victor Vacquier, using magnetic instruments (magnetometres) adapted from airborne devices developed during World War II to detect submarines, began recognizing odd magnetic variations across the ocean floor. This finding, though unexpected, was not entirely surprising because it was known that basalt—the iron-rich, volcanic rock making up the ocean floor—contains a strongly magnetic mineral (magnetite) and can locally distort compass readings. This distortion was recognized by Icelandic mariners as early as the late 18th century. More important, because the presence of magnetite gives the basalt measurable magnetic properties, these newly discovered magnetic variations provided another means to study the deep ocean floor. When newly formed rock cools, such magnetic materials recorded the Earth's magnetic field at the time.

As more and more of the seafloor was mapped during the 1950s, the magnetic variations turned out not to be random or isolated occurrences, but instead revealed recognizable patterns. When these magnetic patterns were mapped over a wide region, the ocean floor showed a zebra-like pattern: one stripe with normal polarity and the adjoining stripe with reversed polarity. The overall pattern, defined by these alternating bands of normally and reversely polarized rock, became known as magnetic striping, and was published by Ron G. Mason and co-workers in 1961, who did not find, though, an explanation for these data in terms of sea floor spreading, like Vine, Matthews and Morley a few years later.

The discovery of magnetic striping called for an explanation. In the early 1960s scientists such as Heezen, Hess and Dietz had begun to theorise that mid-ocean ridges mark structurally weak zones where the ocean floor was being ripped in two lengthwise along the ridge crest. New magma from deep within the Earth rises easily through these weak zones and eventually erupts along the crest of the ridges to create new oceanic crust. This process, at first denominated the "conveyer belt hypothesis" and later called seafloor spreading, operating over many millions of years continues to form new ocean floor all across the 50,000 km-long system of mid–ocean ridges.

Only four years after the maps with the "zebra pattern" of magnetic stripes were published, the link between sea floor spreading and these patterns was correctly placed, independently by Lawrence Morley, and by Fred Vine and Drummond Matthews, in 1963 now called the Vine-Matthews-Morley hypothesis. This hypothesis linked these patterns to geomagnetic reversals and was supported by several lines of evidence:

1. the stripes are symmetrical around the crests of the mid-ocean ridges; at or near the crest of the ridge, the rocks are very young, and they become progressively older away from the ridge crest;
2. the youngest rocks at the ridge crest always have present-day (normal) polarity;
3. stripes of rock parallel to the ridge crest alternate in magnetic polarity (normal-reversed-normal, etc.), suggesting that they were formed during different epochs documenting the (already known from independent studies) normal and reversal episodes of the Earth's magnetic field.

By explaining both the zebra-like magnetic striping and the construction of the mid-ocean ridge system, the seafloor spreading hypothesis (SFS) quickly gained converts and represented another major advance in the development of the plate-tectonics theory. Furthermore, the oceanic crust now came to be appreciated as a natural "tape recording" of the history of the geomagnetic field reversals (GMFR) of the Earth's magnetic field. Today, extensive studies are dedicated to the calibration of the normal-reversal patterns in the oceanic crust on one hand and known timescales derived from the dating of basalt layers in sedimentary sequences (magnetostratigraphy) on the other, to arrive at estimates of past spreading rates and plate reconstructions.

Definition and Refining of the Theory

After all these considerations, Plate Tectonics (or, as it was initially called "New Global Tectonics") became quickly accepted in the scientific world, and numerous papers followed that defined the concepts:

- In 1965, Tuzo Wilson who had been a promotor of the sea floor spreading hypothesis and continental drift from the very beginning added the concept of transform faults to the model, completing the classes of fault types necessary to make the mobility of the plates on the globe work out.
- A symposium on continental drift was held at the Royal Society of London in 1965 which must be regarded as the official start of the acceptance of plate tectonics by the scientific community, and which abstracts are issued as Blacket, Bullard & Runcorn (1965). In this symposium, Edward Bullard and co-workers showed with a computer calculation how the continents along

both sides of the Atlantic would best fit to close the ocean, which became known as the famous "Bullard's Fit".

- In 1966 Wilson published the paper that referred to previous plate tectonic reconstructions, introducing the concept of what is now known as the "Wilson Cycle".
- In 1967, at the American Geophysical Union's meeting, W. Jason Morgan proposed that the Earth's surface consists of 12 rigid plates that move relative to each other.
- Two months later, Xavier Le Pichon published a complete model based on 6 major plates with their relative motions, which marked the final acceptance by the scientific community of plate tectonics.
- In the same year, McKenzie and Parker independently presented a model similar to Morgan's using translations and rotations on a sphere to define the plate motions.

Implications for Biogeography

Continental drift theory helps biogeographers to explain the disjunct biogeographic distribution of present day life found on different continents but having similar ancestors. In particular, it explains the Gondwanan distribution of ratites and the Antarctic flora.

Plate Reconstruction

Reconstruction is used to establish past (and future) plate configurations, helping determine the shape and make-up of ancient supercontinents and providing a basis for paleogeography.

Defining Plate Boundaries

Current plate boundaries are defined by their seismicity. Past plate boundaries within existing plates are identified from a variety of evidence, such as the presence of ophiolites that are indicative of vanished oceans.

Past Plate Motions

Tectonic motion first began around three billion years ago. Various types of quantitative and semi-quantitative information are available to constrain past plate motions. The geometric fit between continents, such as between west Africa and South America is still an important part of plate reconstruction. Magnetic stripe patterns provide a reliable guide to relative plate motions going back into the Jurassic period. The tracks of hotspots give absolute reconstructions, but these are only available back to the Cretaceous. Older reconstructions rely mainly on paleomagnetic pole data, although these only constrain the latitude and rotation, but not the longitude. Combining poles of different ages in a particular plate to produce apparent polar wander paths provides a method for comparing the motions of different plates through time. Additional evidence comes from the distribution of certain sedimentary rock types, faunal provinces shown by particular fossil groups, and the position of orogenic belts.

Formation and Break-up of Continents

The movement of plates has caused the formation and break-up of continents over time, including occasional formation of a supercontinent that contains most or all of the continents. The supercontinent Columbia or Nuna formed during a period of 2,000 to 1,800million years ago and broke up about 1,500 to 1,300million years ago. The supercontinent Rodinia is thought to have formed about 1 billion years ago and to have embodied most or all of Earth's continents, and broken up into eight continents around 600million years ago. The eight continents later re-assembled into another supercontinent called Pangaea; Pangaea broke up into Laurasia (which became North America and Eurasia) and Gondwana (which became the remaining continents). The Himalayas, the world's tallest mountain range, are assumed to have been formed by the collision of two major plates. Before uplift, they were covered by the Tethys Ocean.

Current Plates

Depending on how they are defined, there are usually seven or eight "major" plates: African, Antarctic, Eurasian, North American, South American, Pacific, and Indo-Australian. The latter is sometimes subdivided into the Indian and Australian plates. There are dozens of smaller plates, the seven largest of which are the Arabian, Caribbean, Juan de Fuca, Cocos, Nazca, Philippine Sea and Scotia.

The current motion of the tectonic plates is today determined by remote sensing satellite data sets, calibrated with ground station measurements.

Other Celestial Bodies (Planets, Moons)

The appearance of plate tectonics on terrestrial planets is related to planetary mass, with more massive planets than Earth expected to exhibit plate tectonics. Earth may be a borderline case, owing its tectonic activity to abundant water (silica and water form a deep eutectic.)

Venus

Venus shows no evidence of active plate tectonics. There is debatable evidence of active tectonics in the planet's distant past; however, events taking place since then (such as the plausible and generally accepted hypothesis that the Venusian lithosphere has thickened greatly over the course of several hundred million years) has made constraining the course of its geologic record difficult. However, the numerous well-preserved impact craters have been utilized as a dating method to approximately date the Venusian surface (since there are thus far no known samples of Venusian rock to be dated by more reliable methods). Dates derived are dominantly in the range 500 to 750million years ago, although ages of up to 1,200million years ago have been calculated. This research has led to the fairly well accepted hypothesis that Venus has undergone an essentially complete volcanic resurfacing at least once in its distant past, with the last event taking place approximately within the range of estimated surface ages. While the mechanism of such an impressive thermal event remains a debated issue in Venusian geosciences, some scientists are advocates of processes involving plate motion to some extent.

One explanation for Venus' lack of plate tectonics is that on Venus temperatures are too high for significant water to be present. The Earth's crust is soaked with water, and water plays an important role in the development of shear zones. Plate tectonics requires weak surfaces in the crust along which crustal slices can move, and it may well be that such weakening never took place on Venus because of the absence of water. However, some researchers remain convinced that plate tectonics is or was once active on this planet.

Mars

Mars is considerably smaller than Earth and Venus, and there is evidence for ice on its surface and in its crust.

In the 1990s, it was proposed that Martian Crustal Dichotomy was created by plate tectonic processes. Scientists today disagree, and believe that it was created either by upwelling within the Martian mantle that thickened the crust of the Southern Highlands and formed Tharsis or by a giant impact that excavated the Northern Lowlands.

Valles Marineris is a Tectonic Boundary

Observations made of the magnetic field of Mars by the *Mars Global Surveyor* spacecraft in 1999 showed patterns of magnetic striping discovered on this planet. Some scientists interpreted these as requiring plate tectonic processes, such as seafloor spreading. However, their data fail a "magnetic reversal test", which is used to see if they were formed by flipping polarities of a global magnetic field.

Galilean Satellites of Jupiter

Some of the satellites of Jupiter have features that may be related to plate-tectonic style deformation, although the materials and specific mechanisms may be different from plate-tectonic activity on Earth.

Titan, Moon of Saturn

Titan, the largest moon of Saturn, was reported to show tectonic activity in images taken by the Huygens Probe, which landed on Titan on January 14, 2005.

Exoplanets

On Earth-sized planets, plate tectonics is more likely if there are oceans of water; however, in 2007, two independent teams of researchers came to opposing conclusions about the likelihood of plate tectonics on larger super-earths with one team saying that plate tectonics would be episodic or stagnant and the other team saying that plate tectonics is very likely on super-earths even if the planet is dry.

Remote Sensing in Natural Hazard Assessments

One of the most important tools available to the regional planner is the remote sensing of the environment. Not only is it very useful in the planning process in general, but it is also valuable in detecting and mapping many types of natural hazards when, as is often the case, detailed descriptions of their effects do not exist. If susceptibility to natural hazards can be identified in the early stages of an integrated development planning study, measures can be introduced to reduce the social and economic impacts of potential disasters.

All natural hazards are amenable in some degree to study by remote sensing because nearly all geologic, hydrologic, and atmospheric phenomena that create hazardous situations are recurring events or processes that leave evidence of their previous occurrence. This evidence can be recorded, analyzed, and integrated into the planning process.

Most remote sensing studies concerned with natural hazards have been about an area's vulnerability to a disaster, the monitoring of events which could precipitate a disaster, and the magnitude, extent and duration of a disaster. This chapter tells planners what types of remote sensing information are suitable for identifying and assessing particular natural hazards and where to look for it.

Since the existing remote sensing information may be inadequate for a planning task or phase, this chapter also provides guidelines on selecting and acquiring the appropriate data. Only those sensor systems that are deemed capable of making a insignificant contribution to the development planning process are discussed, with their specific applications to the assessment of each of several natural hazards. It is assumed that planners and other readers are already familiar with basic remote sensing technology and vocabulary. If further details of techniques and/or applications are required, near state-of-the-art information is available in Sabins (1986), Lillesand and Kiefer (1987), and ASP (1983). An excellent overview of satellite imaging systems and disaster management can be found in Richards (1982).

While both aerial and satellite remote sensing techniques are presented, emphasis is placed on satellite-derived sensing because the data provide the synoptic view required by the broad scale of integrated development planning studies. Aerial remote sensing data are useful to natural hazard management for focusing on priority areas, verifying small-scale data interpretations, and providing information about features that are too small for detection by satellite imagery, but extensive aerial surveys commonly exceed the budget constraints of a planning study and may also provide more information than is necessary, particularly during the early stages of the study.

Important Remote Sensing Attributes

Effective utilization of remote sensing data depends on the ability of the user to be accurate and consistent when interpreting photographs, images, graphs, or statistics derived from remote sensing sources. While most planners have been introduced to photo and image interpretation in their formal training, the best use of the data usually requires analysis by people with experience in landform analysis, such as geologists, physical geographers, foresters, etc. A relatively small investment in the services of an experienced interpreter may avoid needless delays and inappropriate use of remote sensing data. Whether or not the planner does his own interpretation, he should have a working knowledge of remote sensing techniques and the capability to assess the validity of an interpretation, as well as the ability to use the derived information.

The factors that determine the utility of remote sensing data in natural hazard assessments are scale, resolution, and tonal or colour contrast. Other factors include area of coverage, frequency, and data cost and availability.

Definitions

Instruments which record electromagnetic radiation emitted or reflected from the earth can be mounted on aircraft or satellites. The former are called aerial or airborne remote sensors and the latter, satellite or spaceborne remote sensors. These instruments record data using optical, electro-optical, optical mechanical, or electronic devices. In this chapter, visual displays analogous to photographs, made with such processes as radar and thermal infrared scanning and produced on a medium other than film, are referred to by the general terms "image" or "imagery"

Scale

The scale to which a photograph or image can be enlarged, with or without optical or computer enhancement, determines in what phase of the development planning study this information should be used. Presentations at scales of 1:500,000 or smaller are useful during the Preliminary Mission and certainly in Phase I, Development Diagnosis, when more detail is not necessary. Imagery at a scale of 1:250,000 or larger is required during the project formulation and feasibility study activities of Phase II when detail is more important and where certain, but less obvious, aspects of natural hazards must be defined. Frequently it is possible to detect natural hazard phenomena on a small scale photograph or image, but it is impossible to annotate it without enlargement to larger scales. Thus, it is necessary to use imagery at scales commensurate with the levels of detail required for the particular stage of the study, as well as the size of the study area itself. In addition, the larger the areal extent of change associated with a natural event, the more useful satellite imagery becomes.

Resolution

Scale is meaningless in the absence of adequate spatial resolution, the capability of distinguishing closely spaced objects on an image or a photograph. Image resolution is determined by the size and number of picture elements or pixels used to form an image. The smaller the pixel size, the greater the resolution. In photography, resolution is limited primarily by the film grain size, but lenses and other technical considerations play important roles.

In both cases, imagery and photography, separability between adjacent features plays a very important part in the identification process. Enlargements of photography or imagery cannot improve resolution but only the working space for the interpretation.

Spectral resolution also needs to be taken into consideration when selecting the type of data since different sensors are designed to cover different spectral regions. Spectral resolution refers to the band range or band width offered by the sensor. Figure depicted earlier, shows the spectral regions most commonly used in remote sensing. Most natural disasters involve spectral changes. Floods lead to significant spectral changes whereas earthquakes lead to little spectral variation due to less spectral contrast in relation to non-affected areas.

Image Contrast

The contrast between features on an image or photograph is a function of the sensor's ability to record the tonal or spectral content of the scene. Different spectral bands of sensing systems may exhibit strong or weak contrasts depending on the regions covered on the electro-magnetic spectrum and the surface viewed. For example, a given band may show little contrast between vegetation types in a forest environment but may show strong contrasts between rock types in an arid area. Hazardous areas such as earthquake fault zones or areas susceptible to landslides may be too small for some sensors, e.g., Advanced Very High Resolution Radiometre (AVHRR) imagery, but may be readily visible on imagery produced by other sensor systems, e.g., Landsat Thematic Mapper (TM). The heavily vegetated and cloud-covered terrain of tropical Latin America and the Caribbean is among the most difficult to interpret geologically, but expert interpreters can detect many natural hazards through physiographic analysis of radar data which can penetrate clouds.

When an image does not provide the detail, resolution, or contrast that is needed, there are several options available. Since the identification of all desired features by interpretation from one sensor is not always possible, a second, completely different type of sensor, or even a combination of sensors, may be needed. Digital data can be enhanced and/or manipulated by using techniques such as contrast stretching, false colour composites, principal component analysis, filtering, and supervised and unsupervised classifications.

Time Frame

The temporal occurrences of natural events will also affect the utility of remotely sensed data. Certain sensors can detect a phenomenon quite readily although their repeat coverage is every 16 days (Landsat). A flood could easily occur and recede within this time frame. On the other hand, desertification of an area can be a long process and the utility of remotely sensed data could be great for monitoring these changes. Events which are seasonal, predictable, or highly correlated with other events are more likely to benefit from imagery than events which occur randomly such as earthquakes or tsunamis.

Remote Sensing Images and Maps

To derive the most benefit from the use of available remote sensing data, the planners should use all supporting information on the study area. Maps are particularly helpful in interpreting remote sensing data. Topographic maps are foremost among maps which help clarify many terrain recognition ambiguities found on remote sensing images. Geological maps bring attention to formations conducive to particular types of hazards. This knowledge can assist in localization and the systematic search for these hazards. Soils maps can serve a similar purpose, but to a lesser extent. Finally, vegetation and land-use maps can provide information on the moisture content, underlying geologic formations, and types of soils present.

In summary, remote sensing imagery should be regarded as data available to assist the planner in the assessment of natural resource and natural hazard information throughout the development of a planning study. The meaning and value of remote sensing data is enhanced through skilled interpretation used in conjunction with conventionally mapped information and ground-collected data.

Output Formats

Output formats consist of different ways in which remote sensing data can be presented. Photographic data are usually used in a film positive format or as a photographic print. Film data and photographic prints can be scanned and converted into digital data by being recorded on a computer compatible tape (CCT). The main advantage of digital data is the fact that they can be quantified and manipulated using various image processing techniques. Satellite or other images recorded on a CCT can be presented in a film positive format or photographed directly from the display monitor.

Aerial Remote Sensing

Aerial remote sensing is the process of recording information, such as photographs and images, from sensors on aircraft. Available airborne systems include aerial cameras, multispectral scanners, thermal infrared (IR) scanners, passive microwave imaging radiometres, and side-looking airborne radars (SLAR). The systems offering the most practical and useful data in the context of integrated development planning and natural hazard assessments are aerial cameras, multispectral scanners, and thermal IR scanners and SLAR. This section describes the characteristics of the photography or imagery obtained from these three systems.

Availability of aerial remote sensing imagery varies for the type of data required. Aerial photography is readily available for many areas of study in most parts of the world, although in some instances it must be declassified for non-military use by the government of the country involved in the study. Radar imagery is also frequently classified. Acquisition of infrared (IR) and radar data is more complex than aerial photography, although for a large area, radar may be less expensive than photography. Due to the specialized systems and operators required to produce IR and SLAR imagery, such data are usually available only from a limited number of organizations which either own or lease the systems. The cost to mobilize aircraft, equipment, and crews is high, but the cost of data coverage per line kilometre or per unit area can be reasonable if the area to be flown is large.

In addition to the type, availability, and cost of data, the planner should consider the conditions under which the acquisition of the appropriate data is taking place. Each sensor type has an optimum time of day, season, and/or table of appropriate conditions under which the best results are obtained. Also, to establish the current status of a hazard such as a volcano's activity, the interpretation of thermal IR imagery must be made close to the time of acquisition, and anomalies should be checked out immediately to determine the magnitude of temperatures that correlate with them. Currently obtained data, flown under similar instrument, weather, and terrain conditions, may be used to compare temporal variations of the hazard. In this way thermal pattern changes may be determined.

Thermal IR imagery information is the most transitory of any sensor data. There is a procession of changes in the thermal contrasts between the different materials on the ground, both terrain and vegetation. The transitions occur over daily and seasonal cycles and are modified considerably by the weather, soil, climate, relief, slope direction, and land-use practices. In spite of these masking variations, the thermal contrasts resulting from volcanic and geothermal activity can be interpreted by an experienced thermal IR interpreter.

The primary utility of SLAR imagery is in the interpretation of the relatively unchangeable elements of basic geologic structure and geomorphologic conditions. As a result, it is useful in studying many features related to natural hazards. Special SLAR image data acquisition is not normally feasible in a planning study budget, but previous coverage of the study area may be available. If it does exist, it should be sought and used to its fullest extent.

Both IR and SLAR imagery can be used in a stereoscopic mode but only where adjacent flight lines overlay. Since distortion due to air turbulence and/or differential altitude occurs during the raster-like development of each image as the aircraft moves forward, the stereoscopic model is imperfect. Despite these distortions the stereoscopic dimension is definitely an asset in helping to define natural hazards.

Aerial Photography

Of all the sensors, aerial photography gives the closest representation to what the human eye sees in terms of wavelength response, resolution, perspective, stereoscopic viewing, and tonal or colour values. The interpreter familiar with photographs can easily interpret these scenes, whereas other sensors, such as thermal IR scanners and SLAR systems, produce imagery whose appearance and physical basis is completely foreign to the inexperienced eye. Aerial photographs are probably the remote sensing data source with which the planner is most familiar (OAS, 1969).

Scales and Wavelengths

The most useful scales for aerial photographs range from 1:5,000 to 1:120,000. The need for reconnaissance type of information over large areas limits the use of photographs to the scales of 1:40,000 or less.

Photography is limited to the optical wavelengths which are composed of ultraviolet (UV), visible, and near-IR portions of the electromagnetic spectrum. The first and last of these portions are recoverable on film under special film-filter conditions. The near-IR wavelengths are the reflective part of the larger infrared portion, which also includes emitted or thermal wavelengths.

Type of Film

Aerial photography may be obtained using black and white film, the least expensive medium, or with conventional colour or colour IR film. The type of film that should be used depends on its utility for a particular terrain being studied and the cost of the film. The speed of the film is also an important factor: the slower colour films may not be used where the terrain is too dark, such as areas of ubiquitous heavy vegetation or predominantly dark rocks.

The two general types of black and white films used most frequently are the panchromatic and IR-sensitive films. Panchromatic films, which are negative materials having the same approximate range of light sensitivity as the human eye, are regarded as the standard film for aerial photography. It is the least expensive medium for aerial mapping and photo interpretation, but it may not be the logical choice for a given study area.

Black and white IR-sensitive film, although not commonly used, is a better choice for the penetration of strong haze and/or lush vegetation in humid tropical areas. It renders surface water, moisture, and vegetation contrasts much better than the standard film, and, as a result, can be an effective tool in regional planning and natural hazard assessments in humid tropical areas. There is, however, a diminution of detail in shadowed areas since scattered cooler light (blue end) is filtered out.

In high relief areas, it is best to shoot close to midday using IR films. In areas of low relief, photographs should be taken when the sun is low on the horizon (10° to 30°), causing shadows on the fine-textured surface. Low-sun-angle photography (LSAP) emphasizes textural characteristics of particular rock types, discontinuities, and the linear topographic features associated with faults and fractures. Vegetation types, both natural and cultivated, can also be defined to a large extent on a textural basis, and this may provide further information on the terrain. Almost any state-of-the-art aerial camera can capture LSAP using panchromatic or red-filtered infrared film.

The use of colour films for natural hazard assessment takes various forms: negative film from which positive colour prints are made, and positive transparencies, including colour slides. To a limited extent, the negative films can be printed to emphasize certain colours and offer the ease of handling of prints. They do not have, however, the sharpness and dynamic colour range of the positive transparencies, which are significantly better for interpretation purposes.

There are two major spectral types of colour film: the natural or conventional colour film, which covers the visible spectrum, and colour IR film (green through near IR). The former is available as a negative (print) film and positive transparency, and the latter is available only as a positive transparency.

The IR colour film response is superior to that of natural colour films for a number of reasons. First, the yellow filter required for its proper use eliminates blue light that is preferentially scattered by the atmosphere. Eliminating much of the scattering greatly improves the contrast. Second, the differences in reflectance within vegetation types, soils, and rocks are commonly greater in the photographic IR component of this film. Third, the absorption of infrared and much of the red wavelengths by water enables a clearer definition of bodies of water and areas of moisture content. And fourth, the diminution of scattered light in shadowed areas enhances relief detail, thus improving the interpretation of the geomorphology. In view of these attributes, colour IR film is preferred if colour aerial photography is desired for humid tropical climates.

Radar

Radar differs from aerial photography as an aerial remote sensor. Unlike photography, which is a passive sensor system using the natural reflection from the sun, radar is an active sensor that produces its own illumination. Radar illuminates the terrain and then receives and arranges these reflective signals into an image that can be evaluated. These images appear similar to black and white photographs. The best use of airborne radar imagery in the development planning process and natural hazard assessments is the identification of geologic and geomorphologic characteristics. Radar imagery, like photography, presents variations in tone, texture, shape, and pattern that signify variations in surface features and structures. Of these elements, tonal variations which occur in conventional aerial photographs are the same as the eye sees. The tonal variations, which occur in radar images and appear as unfamiliar properties, are the result of the interaction of the radar signal with the terrain and vegetation. Just as it is not essential to fully understand the optical theory and processes involved with photography to be able to use aerial photographs, it is also possible to use radar images without a thorough understanding of electromagnetic radiation.

However, an interpreter needs to know something about how the image is formed in order to interpret it correctly and to appreciate fully the potential and limitations of radar. A skilled interpreter need only become familiar with the parametres that control radar return, understand their effect on the return signal, and recognize the effect of the side-looking configuration of the sensor on the geometry of the return signal.

Many useful radar images have been acquired in X-band, K-band, and Ka-band wavelengths. However, X-band airborne radar systems are currently the most commonly offered by commercial contractors. In this band-width there are two basic types of systems: real aperture radar (RAR) and synthetic aperture radar (SAR). Real aperture or "brute force" radar uses an antenna of the maximum practical length to produce a narrow angular beam width in the azimuth (flight line) direction. The longer the antenna, the narrower and the azimuth beam. A typical length is 4.5m, which approaches a maximum practical size for aircraft. For this reason the SAP was developed. The SAR is capable of achieving higher resolution without a physically large antenna through complex electronic processing of the radar signal.

The resulting resolution, coupled with the small scales at which images can be acquired, makes radar more suitable than photographic surveys for covering large areas. While RAR has a simple design and does not require sophisticated data recording and processing, its resolution in the range direction is relatively limited in comparison with the SAR of the same waveband. SAR maintains its high resolution in the range direction at long distances as well as its azimuth resolution. Resolution with SAR approaches 10m in azimuth and range.

Thermal Infrared Scanners

An airborne electro-optical scanner using a semiconductor detector sensitive to the thermal IR part of the spectrum is the best way to produce imagery that defines the thermal pattern of the terrain. Alternative methods using a television-like presentation have inadequate spatial resolution and thus cannot be used effectively from aircraft altitudes. They also lack adequate thermal resolution.

Table: *Radar Wavelength And Frequencies Used In Remote Sensing For Aircraft Radar Systems*

Band Designation [a/]	*Wavelength (cm)*	*Frequency(v), GHz (10^9 cycles/sec^{-1})*
Ka (0.86cm)	0.8 to 1.1	40.0 to 26.5
K	1.1 to 1.7	26.5 to 18.0
Ku	1.7 to 2.4	18.0 to 12.5
X (3.0cm, 3.2cm)	2.4 to 3.8	12.5 to 8.0
C	3.8 to 7.5	8.0 to 4.0
S	7.5 to 15.0	4.0 to 2.0
L (23.5cm, 25.0cm)	15.0 to 30.0	2.0 to 1.0
P	30.0 to 100.0	1.0 to 0.3

[a/]Wavelengths commonly used in imaging radars in imaging radars are shown in parentheses

Source: Sabins, Floyd F., Jr. Remote Sensing: Principles and Interpretation (New York: W.H. Freeman, 1986).

Spatial resolution in scanners decreases with altitude above the terrain. Most commercial thermal infrared systems have spatial resolutions which provide for 2m to 2.5m resolution per 1,000m altitude at the nadir point (the point in the ground vertically below the camera) of the scan. Increasing the altitude above terrain to 2,000m would produce 4m to 5m spatial resolution.

Commonly, the 3.0 mm to 5.5 mm band provides the best information for "hot" objects (active volcanic vents, hot springs, etc.), while the 8.0 mm to 14.0 mm band provides the best information for features that are at ambient or cooler temperatures (flooding streams under canopies, warm springs, etc.). Frequently in studies involving IR surveys both bands are used to provide simultaneous imagery.

Properties of the airborne IR scanner system indicate that its practical use is restricted to the lower altitudes (under 3,000m) and, consequently, to relatively smaller areas than either radar or aerial photography. In natural hazard assessments, its best use would be in areas that are known or suspected to be areas of volcanism or where abnormal moisture conditions indicate dangerous situations. The latter may include, for example, trapping of water along active faults, or in back of landslide slumps, or moisture conditions associated with karst terrain. IR scanning systems have drawbacks, but their unique capability of thermal imaging is unsurpassed. In addition, they can provide critical information for relatively small areas once specific hazard-prone areas have been identified.

Advantages and Limitations of Photography, Radar, and Thermal IR Scanners

Photography and Radar: Both aerial photography and radar have advantages and limitations. Photography cannot be used at any time in any weather as can radar. Radar can map thousands of square miles per hour at geometric accuracies conforming to national mapping standards. An area can be surveyed much more rapidly by radar than by aerial photography, and the final product provides an excellent synoptic view. Distance can be measured more accurately on radar than photography, and maps as large as 1:24,000 scale have been produced experimentally. The RADAM project of Brazil covered the country completely at a scale of 1:250,000. On the other hand, photography at the same scale shows considerably more detail, and it provides an excellent stereoscopic model for interpretation purposes in contrast to a more limited, but still useful, model obtained from radar. Aerial photography has the advantage of offering instantaneous scene exposures, superior resolution, ease of handling, and stereoscopic capability.

Thermal IR Scanners

Airborne electro-optical scanners, in general, can cover the electromagnetic spectrum using semiconductor electronic sensors from the UV through the visible and near IR into the thermal IR range of the spectrum. The utility of the UV spectrum in natural hazard and resource investigations has yet to be demonstrated, particularly when the image is degraded due to intense scattering of its rays. Scanners in the visible range are useful, especially when two or more wavebands are algebraically combined or manipulated.

Scanning imagery, because of its technique of recording a raster on film or tape, produces inherent distortions in the final built-up image scene. The lateral distortion from the flight line is reasonably corrected in the scanner system. Along the flight line, however, the rapid changes of altitude above the terrain during the formation of one scene from many scan lines produces many distortions. The persistent movement of the aircraft on three axes with limited stabilization presents the same problem. These distortions result in images that are difficult to interpret and whose location is difficult to identify, especially in mountainous and/or forested terrains. Despite these deficiencies, scanning from aircraft continues to be a very valuable method of obtaining thermal infrared imagery with reasonable spatial and thermal resolution.

In summary, aerial remote sensing provides information from aerial photographic cameras, side-looking radar, and thermal imaging scanners that is unsurpassed in resolution in their respective coverage within the electromagnetic spectrum. These systems produce imagery that ranges from the familiar visible spectrum to the unfamiliar infrared and microwave radar (short radio) spectra. This information can be used in conjunction with conventional maps of all kinds to enhance the data available to the planner.

Satellite Remote Sensing

This section describes several satellite remote sensing systems which can be used to integrate natural hazard assessments into development planning studies. These systems are: Landsat, SPOT satellite (Système Probatoire pour l'Observation de la Terre), satellite radar systems, Advanced Very High Resolution Radiometre (AVHRR) on NOAA-10 and 11 satellites, metric camera, large format camera (LFC), and Sojuzkarta. Remote sensing from satellite vehicles has become increasingly important following the successful launch of the Landsat 1 satellite (formerly ERTS-1) in 1972. Since then many satellites with remote sensing capabilities have been developed and used successfully.

The Landsat multispectral scanner (MSS) provided the first practical imagery in four different spectral bands from space. The characteristics of this and other Landsat sensors are summarized in Figures. The accompanying return beam vidicon (RBV) sensor on this and subsequent satellites of this series were never noticed by scientists and planners like the MSS. The broad areal coverage of the Landsat sensors and the others that have followed, together with the capability to process the sensor data digitally, has made the satellite-derived data useful to regional planners and others interested in natural hazard assessments. The synoptic view of lands targeted for development can be imaged in an instant of time, Satellite imagery can provide a continuity of viewing conditions for large areas that is not possible on aerial photographic mosaics.

In addition to MSS, other satellite-borne sensors warrant discussion since they are potential tools for assessing natural hazards. Each sensor has its advantages and limitations in coverage of areas of interest and its resolution capability to define certain types of hazards. Some sensors are experimental, and provide limited areal coverage and lack temporal continuity. However, where coverage is available for a study area, the sensor data should be used in conjunction with existing data derived from Landsat or SPOT. The data derived can produce an inexpensive synergistic effect by combining data from more than one part of the spectrum, and are well worth the relatively small expense. Ideally, it would be desirable to use a "multi-stage" approach to the natural resource and natural hazard assessments. This would involve using aerial photography and ground checks to establish more detailed knowledge of sample or representational sites. This may then be extrapolated over a larger area using Landsat or other satellite-derived data. Figure given shows the needed imagery characteristics for the assessment of various natural hazards-earthquakes, volcanic eruptions, landslides, tsunamis, desertification, floods, and hurricanes-for purposes of planning and mitigation.

Landsat

Since the Landsat series of satellites have been operational for a long period of time, there is a very large data base available, both in areal coverage and in repetitive coverage, through different seasons and during periods of natural disasters. Landsat MSS coverage exists from 1972 to the present in four spectral bands at 80m resolution. The thematic mapper (TM) was introduced on Landsat 4 in 1982, with seven spectral bands, six of them with 30m resolution and one in the thermal IR range with 120m resolution.

Sensor data are digitally transmitted to ground stations in various parts of the world where they are recorded on magnetic tapes and preprocessed to improve their radiometric, atmospheric, and geometric fidelity. Ground receiving stations that cover Latin America and the Caribbean are in California, Maryland, Brazil, and Argentina.

Although none of the existing satellites and their sensors has been designed solely for the purpose of observing natural hazards, the variety of spectral bands in visible and near IR range of the Landsat MSS, and TM and the SPOT HRV sensors provide adequate spectral coverage and allow computer enhancement of the data for this purpose. Repetitive or multitemporal coverage is justified on the basis of the need to study various dynamic phenomena whose changes can be identified over time. These include natural hazard events, changing land use-patterns, and hydrologic and geologic aspects of a study area.

The use of Landsat MSS and TM imagery in natural resource evaluations and natural hazard assessments is facilitated by the temporal aspect of available imagery. Temporal composites from two or more different image dates allow the recognition of hazard-related features that have changed, such as alteration of floodplains or stream channels and large debris slides, and to some extent, early recognition of disasters that evolve over time, such as drought or desertification. Chapter 8 has a detailed discussion of the use of Landsat sensors in flood hazard assessments. Specific manipulation and combination of the MSS or TM tape data of various bands of the same scene can increase the utility of the data.

Three-dimensional analysis, or stereoscopy, is essentially missing on the MSS and TM data. With MSS on Landsat 1, 2, and 3, there is an 18-day cycle, and sidelap is 14 percent at the equator, increasing poleward to 34 percent at latitude 40 and to 70 percent at polar latitudes. (Sidelap is the amount of overlapping of adjacent image coverage.) On Landsat 4 and 5, a lower altitude and a 16-day cycle with wider spacing results in only 7.6 percent sidelap at the equator and negligible increase towards the poles for both MSS and TM data. Unfortunately, the areas at the lower latitudes where our interests lie have the minimal stereoscopic coverage.

***Table:** Characteristics Of Landsat Sensors*

Sensor [a]	*Landsat Platform*	*Spectral bands and range (micrometers)*		*Altitude (km)*	*Resolution (m)*	*Image size (km)*	*Coverage repeat*
RBV	1,2,3	PAN	0.505-0.750	920	79x59 [d]	185x185 [d]	every 18 days
		1 [b]	0.475-0.575		30x30 [e]	99x99 [e]	
		2 [c]	0.580-0.680				
		3 [c]	0.690-0.830				
MSS	1,2,3,4,5	4 (green)	0.5-0.6 [f]	920 [g]	79x57 [g]	185x185 [f]	every 18 days [g]
		5 (red)	0.6-0.7	705[h]	60x60[h]	185x170 [g]	every 16 days [h]
		6 (near IR)	0.7-0.8		237x237 [i]		
		7 (near IR)	0.8-1.1				
		8 (thermal)	10.4-12.6 [b]				
TM	4,5	1	0.45-0.52	705	28.5x28.5 [k]	85x170	every 16 days
		2	0.52-0.60		120x120 [l]		
		3	0.63-0.69				
		4	0.76-0.90				
		5	1.55-1.75				
		6	10.40-12.50 [j]				
		7	2.08-2.35				

[a] RVB, Return Beam Vidicon; MSS, Multispectral Scanner; TM, Thematic Mapper; IR, Infrared.

[b] Panchromatic and Landsat 3 only

[c] Bands 1,2,3 on Landsat 1 and 2 only

[d] Landsat 1 and 2

[e] Landsat 3

[f] Also called bands 1 to 4 on Landsat 4 and 5

[g] Landsat 1 to 3

[h] Landsat 4 and 5

[i] Band 8 on Landsat 3

[j] Thermal

[k] Bands 1 to 5 and 7

[l] Band 6 only

Source: Adapted from Budge, T. A Directory of Major Sensors and Their Parametres (Albuquerque, New Mexico: Technology Application Centre, University of New Mexico, 1988).

If the terrain is flat with little relief, the little stereoscopic sidelap present would not be effective. In areas of rugged terrain, any small stereoscopic coverage would be welcome, especially if it fell within a critical part of a project area. The return beam vidicon (RBV) is a framing camera system that operates as an instant television camera. It has not enjoyed the same popularity as the MSS, even though it provides useful information. Landsat 1 and 2 carried three RBVs that recorded green, red, and solar IR images of the same scene as the MSS did. They were capable of producing colour IR images of 80m resolution, the same as the MSS, but were decidedly inferior due to technical problems. Landsat 3 carried an RBV system that acquired single black and white images in quadrants of the MSS scene in the 0.50mm to 0.75mm waveband, a spectral response from the green through red. The ground resolution was 40m, much better than the existing MSS and the earlier RBV resolution, enabling the recognition of natural hazard evidence of smaller scale.

The broad response of the RBV, however, did not enhance any particular feature or differentiate vegetation or rocks as well as the MSS bands. Its advantage lay primarily in providing higher spatial resolution for larger scale mapping of spectrally detectable features. In this regard it complemented the lower resolution MSS data which covered the same area. The RBV system was dispensed with entirely on Landsat 4 and 5, leaving only the MSS and TM sensors. The former was included to continue the temporal library with that type of sensor data and their 80m spatial resolution. The TM, with a 30m resolution, negated any requirement for the ineffective and little-used RBV system. Despite its absence on Landsat 4 and 5, RBV data of certain heavily vegetated tropical areas may be the only available source of data with adequate resolution for temporal comparison with later TM data.

Sources for Landsat Imagery

Argentina:

Established: November 1980 Reception: MSS Distribution Centre: Comisión Nacional de Investigaciones Espaciales Centro de Procesamiento Avenida Dorrego 40101425 Buenos Aires, Argentina Telephone: 722-5108; Telex: 17511 LANBAAR

Brazil

Established: May 1974 Reception and Processing: MSS and TM Distribution Centre:INPE-DGICaixa Postal 01, Cachoeira Paulista, SPCEP 12630, Sao Paulo, Brazil Telephone: (125) 611507; PBX: (125) 611377 Telex: 1233562 INPE BR United States Established: July 1974 Reception and Processing: MSS and TM Distribution Centre:

Earth Observation Satellite Company (EOSAT) 4300 Forbes Blvd., Lanham, Maryland 20706Telephone: (301) 552-0500 or 800-344-9933 Telex: 277685 LSAAT UR.

The thermal IR portion of the TM was originally placed in the 10.4 mm to 12.5 mm spectral window where the earth's radiant energy is so low that a large detector is required. This resulted in a 120m ground resolution cell which generalized thermal detail, limiting its value for detecting the subtle and finely detailed geothermal changes associated with volcanic activity. The thermal resolution is 0.5° K (degrees Kelvin), which by airborne IR scanner standards (0.1° K or smaller), is poor. The best possible application in natural hazard assessments may be in active floodplain delineation and perhaps as a crude index to regional volcanic activity. The thermal infrared band (band 8) in Landsat 3 (10.4 mm - 12.5; mm with 240m spatial resolution) never worked properly and is, therefore, of no consequence for the applications discussed here. The blue-green band (0.45 mm - 0.52 mm) in the TM system (band 1) is unique among sensors aboard natural/ resource-oriented satellites. The reason that this band has not been a part of the spectrum sought from satellites is the severe scattering of the blue light, which can badly degrade the image contrast where there is high humidity and/or high aerosol content in the atmosphere. However, in water, blue light has the best penetration capability in the visible spectrum. In clear, sediment-free waters it can define sea bottoms to depths of 30 or more metres, depending primarily on the angle of incidence of the sun's illumination and the reflectance of the bottom. This property is useful for determining offshore slope conditions relevant to potential tsunami run-up.

***Table:** Satellite remote sensing: approximate costs of basic data acquisition (CCTs, June 1989)*

System	***Cost per km²***
Landsat MSS	US$.02
Landsat TM	US$.11

***Table:** Characteristics Of Spot Sensors*

SPOT SENSOR: MULTISPECTRAL HIGH-RESOLUTION VISIBLE (HRV)			
Band	**Wavelength (? m)**	**Resolution (m)**	**Image Format**
XS1	0.50-0.59	20	60km swath with vertical
XS2	0.61-0.68	20	viewing angle and up to
XS3	0.79-0.89	20	80km with +/- 27° viewing angle from vertical
SPOT SENSOR: PANCHROMATIC HIGH-RESOLUTION VISIBLE (HRV-P)			
Band	**Wavelength (? m)**	**Resolution (m)**	**Image Format**
P	0.51-0.73	10	60km swath with vertical viewing angle and up to 80km with +/- 27° viewing angle from vertical

Spot

The SPOT satellite with its High Resolution Visible (HRV) sensors is similar in many respects to the Landsat satellite with its MSS and TM sensors. The HRV multispectral sensor (XS) ranges from the green wavelength into the near IR. The HRV-XS coverage is in three spectral bands rather than the four found in the MSS, but with much higher spatial resolution (20m versus 80m), although it covers only about 1/9 of the area covered by a Landsat scene. Additionally, SPOT carries a panchromatic sensor (HRV-P) which covers the green through red portions of the visible spectrum in a single band with 10m resolution. Both HRV sensors cover a 60km swath along the orbit path. It is possible to obtain simultaneous side-by-side coverage from each sensor, producing a 117km swath width, although this capability has not been frequently used.

The SPOT sensors have the unique capability of being pointable, 27 degrees to the left or right of the orbital track. This feature allows for repeated off-nadir viewing of the same ground swath, producing image stereopairs. The base-height ratios range from 0.75 at the equator to 0.50 at the mid-latitudes. This provides a strong vertical exaggeration. This third dimension, if it is available for a particular study area, together with the higher image resolution, can make SPOT'S sensors superior to those of Landsat if greater spectral resolution is not required.

Satellite Radar Systems

There is considerable radar coverage throughout the world, and more space-derived radar data can be expected in the future.

The family of space radars stems from the Seasat (U.S.A.) radar, which was a synthetic aperture system that was especially designed for studying the ocean surface. In this capacity it had a large (70° average) depression angle to study the relatively flat ocean surface. For this reason Seasat's usefulness for imaging land extended to those land areas of low relief. During its short life in 1978, it managed to acquire a large amount of data from Western Europe, North and Central America, and the Caribbean.

Seasat was followed by Space Shuttle imaging radars SIR-A and SIR-B. Data from these radars was obtained from Space Shuttle flights in 1981 and 1984. Their characteristics along with those of Seasat are shown. SIR-A and SIR-B provided greater worldwide coverage, including large parts of Latin America, because the image data were recorded on board the Space Shuttle rather than telemetreed to a limited number of receiving stations within range of the spacecraft, as was the case with the unmanned Seasat radar satellite.

Table: *Characteristics Of Seasat, Sir-a, And Sir-b Systems*

Characteristics	*Seasat (1978)*	*SIR-A (1981)*	*SIR-B (1984)*
Repeat Coverage	irregular, northern hemisphere	little to none	little to none
Resolution	25x25m	40x40m	25x(17-58)m
Wavelength (23.5cm)	L-band	L-band	L-band
Latitude coverage	72°N-72°S	50°N-35°S	58°N-58°S
Altitude	790km	250km	225km
Image-swath width	100km	50km	40km

Source: Adapted from Budge, T. A Directory of Major Sensors and Their Parametres (Albuquerque, New Mexico: Technology Application Centre, 1988).

Sources of Satellite Radar Imagery

SIR-A and SIR-B: National Space Science Centre World Data Centre A for Rockets and Satellites Code 601NASA/Goddard Space Flight Centre Greenbelt, Maryland 20771, U.S.A. Telephone: (301)286-6695.

Seasat: NOAA, National Environmental Satellite Data and Information Service World Weather Building, Room 100 Washington, D.C. 20233, U.S.A. Telephone: (301)763-8111

The long wavelengths of these radar systems permit potential subsurface penetration between 2m and 3m in extremely dry sand (Schaber etal., 1986). There may be areas of hyperacidity in South America that may permit this type of penetration. This property may have some application to natural hazard assessment that is not readily apparent, as well as to integrated development planning studies. The problem seems to be that while significant amount of radar coverage is available, much has yet to be acquired where needed.

The SIR series of radar data acquisition is expected to continue with SIR-C in the future. Other radar sensors will be placed into orbit soon: Canada's Radarsat, a C-band (6.0cm) radar designed to provide worldwide stereoscopic coverage, is planned for the 1990s; the European Space Agency expects to launch a C-band synthetic aperture radar aboard the Earth Resources Satellite (ERS) in 1990; and Japan will launch an L-band imaging radar satellite in 1991. Thus, it can be expected that more radar imagery is forthcoming which will provide additional tools for natural hazard assessment.

AVHRR

The Advanced Very High Resolution Radiometre (AVHRR) on board the NOAA-7 through 11 satellites would not normally be considered useful for natural hazard assessments on the basis of its low resolution (1.1 km at nadir) alone. However, its large swath width of 2253km provides daily (day and night) coverage of the inhabited parts of the earth. The near-nadir viewing repeat cycle is nine days, but the same area is still viewable from different angles within the swath from space. This results in complicated radiometric and geometric comparisons between different dates of data acquisition.

This scanning radiometre has 5 bands, which include band 1 (green to red), band 2 (red to reflected IR), band 3 (middle IR), band 4 (thermal IR), and band 5. The most useful bands are the thermal IR bands 4 and 5 particularly where moisture or ice is involved.

These have been successfully used to delineate flooded areas using temporal analysis techniques within 48 hours following a major flood event (Wiesnet and Deutsch, 1986). The thermal resolution in these bands is better than the Landsat TM thermal band 6 but with a significant trade-off loss in spatial resolution (1.1km versus 120m, respectively).

Metric Camera

The metric camera was an experiment on the STS-9/Spacelab 1 Mission in 1983 to determine whether topographic and thematic maps at medium scales (1:50,000 to 1:250,000) could be compiled from mapping camera images taken from orbital altitudes. Due to a late November launch date, illumination conditions were poor over many of the candidate target areas. As a result, slower shutter speeds had to be used than were planned, causing some image smear. Nevertheless, high quality images with a photographic ground resolution of about 20m were obtained on a 23cm x 23cm format panchromatic and colour IR film. Analysis has shown that these images may be used for mapping at a scale of 1:100,000. On this mission, despite the fact that many problems were encountered, an area of more than 11 million km^2 was covered. Plans are now underway to modify the camera to compensate for forward image motion and to refly it. It is expected that a ground resolution of about 10m would be obtained, permitting maps with a scale as large as 1:50,000 (Schroeder, 1986).

Table: *Avhrr Characteristics*

Platform: NOAA satellites (formerly Tiros)			
Spectral bands	*Tiros-N*	*NOAA-6.8,10*	*NOAA-7,9,d,H,I,J*
1	0.55 - 0.90	0.58 - 0.68	0.58 - 0.68
2	0.725 - 1.00	0.725 - 1.00	0.725 - 1.00
3	3.55 - 3.93	3.55 - 3.93	3.55 - 3.93
4	10.50 - 11.50	10.50 - 11.50	10.30 - 11.30
5	none	none	11.50 - 12.50
Altitude: 833-870km			
Resolution:	Large Area Coverage (LAC): 1km		
	Global Area Coverage (GAC): 4km		
Image size: 2253 km swath			
Repeat coverage: Daily, worldwide			

Source of Avhrr Imagery

Satellite Data Services DivisionNOAA/NESDIS/NCDCWorld Weather Building, Room 100 Washington, D.C. 20233, U.S.A. Telephone: (301) 763-8111.

Source of Metric Camera Photography

Deutsche Forschungs- und Versuchsanstalt furLuft- und Raumfahrt e.V. (DFVLR) OberpfaffenhofenD-8031 Wessling Federal Republic of Germany. Ground coverage of 190km x 190km per photograph frame was obtained using a 305mm lens from an altitude of 250km, yielding an image scale of 1:820,000. Overlap of 60 to 80 percent, obtained for topographic mapping purposes, is of great value in interpretation for natural hazards. The high resolution and stereocoverage make this photographic sensor system a potentially useful tool when sufficiently enlarged.

Five lines of metric camera photography cover parts of Latin America, and with resumption of the U.S. Space Shuttle programme, additional high quality space photographs of areas of interest may become available.

Large Format Camera

The large format camera (LFC) photography was obtained during a Space Shuttle flight in October 1984. The term "large format" refers to the 23cm by 46cm film size, which was oriented with the longer dimension in the line of flight. LFC acquired 1,520 black and white, 320 normal colour, and 320 IR colour photographs, covering many areas within Latin America and the Caribbean. The scale of the photographs ranges from 1:213,000 to 1:783,000 depending on the altitude of the Space Shuttle, which varied between 239km and 370km. The swath covered a range between 179km and 277km, and each frame covered between 360km and 558km in the flight direction. Forward overlap up to 80 percent was obtained, allowing vertical exaggerations of 2.0, 4.0, 6.0, and 7.8 times in the stereomodels. Most photographs were taken with 60 percent overlap, which provided 4 times vertical exaggeration and an excellent stereomodel. The spatial resolution was about 3m for the black and white film and about 10m for the colour IR film.

The availability of this excellent stereophotography, which can be enlarged ten times or more with little loss of image quality, is limited to certain areas covered by the ground track of the Space Shuttle. Some of this coverage includes clouds or heavy haze, but despite the limitations of coverage and occasional poor quality, the existing photography should be examined for its possible use in any regional natural hazards assessment and planning study. Given the range of tools available for aerial and satellite remote sensing, their applications vary based on the advantages and limitations of each. The planner can regard each of these as a potential source of information to enhance natural resource evaluation and natural hazard assessment. The next section covers some of the applications of photographs and images in natural hazard assessments.

Sojuzkarta

Sojuzkarta satellite data consist of photographs taken with the KFA-1000 and KM-4 camera. Computer compatible tapes (CCTs) for digital image processing are not available, although it is possible to convert the data into digital format by using a scanner. Photographs obtained through the KFK-1000 camera have 5m resolution in the panchromatic mode and 10m resolution in the colour mode; scales range from 1:220,000 to 1:280,000. KM-4 photography has a 6m resolution and is available at scales of 1:650,000 and 1:1,500,000. Applications of this sensor to natural hazard studies are likely to be in desertification monitoring, flood hazard and floodplain, and landslides studies.

Applications of Remote Sensing Technology to Natural Hazard Assessments

For purposes of assessing natural hazards in the context of integrated development planning studies, it is not necessary to have real-time or near real-time remote sensing imagery. What is required is the ability to define areas of potential exposure to natural hazards by identifying their occurrence or conditions under which they are likely to occur and to identify mechanisms to prevent or mitigate the effects of those hazards. This section considers the practical detectability by remote sensing technology of the potential for floods, hurricanes, earthquakes, volcanic eruptions and related hazards, and landslides. It will become evident that some of these hazards are interrelated, e.g., floods and hurricanes; earthquakes, volcanoes and landslides.

The ability to identify these natural hazards or their potential for occurring depends on the resolution of the image, the acquisition scale of the sensor data, the working scale, scenes free of clouds and heavy haze, and adequate textural and tonal or colour contrast. The availability of stereomodels of the scene being studied can greatly enhance the interpretation.

After a hazard is identified, formulating appropriate mitigation measures and developing response plans may require different remote sensing data sets. This additional remote sensing data needed are most likely to include greater detail of the infrastructure, e.g., roads and facilities. This may have to be derived from aerial photography.

Floods

Floods are the most common of natural hazards that can affect people, infrastructure, and the natural environment. They can occur in many ways and in many environments. Riverine floods, the most prevalent, are due to heavy, prolonged rainfall, rapid snowmelt in upstream watersheds, or the regular spring thaw. Other floods are caused by extremely heavy rainfall occurring over a short period in relatively flat terrain, the backup of estuaries due to high tides coinciding with storm surges, dam failures, dam overtopping due to landslides into a reservoir, and seiche and wind tide effects in large lakes. Occasionally an eruption on a glacier or snow-covered volcanic peak can cause a flood or a mudflow in which the terrain is radically changed and any agrarian development is totally destroyed, frequently with much loss of life. It is impossible to define the entire flood potential in a given area. However, given the best remote sensing data for the situation and a competent interpreter, the evidence for potential flood situations can be found or inferred. The most obvious evidence of a major flood potential, outside of historical evidence, is identification of floodplain or flood-prone areas which are generally recognizable on remote sensing imagery. The most valuable application of remote sensing to flood hazard assessments, then, is in the mapping of areas susceptible to flooding.

Synoptic satellite sensor coverage of a planning study area is the practical alternative to aerial photography because of cost and time factors. The application of Landsat MSS imagery to floodplain or flood-prone area delineation has already been demonstrated by comparing pre-flood scenes with scenes obtained at the height of the flood, using Landsat MSS band 7 (near IR) images in a colour additive viewer (Deutsch etal., 1973). This temporal comparison can now be done pixel by pixel in a computer. Landsat TM, with its greater spatial resolution than MSS data (30m versus 80m) and its additional spectral coverage (7 bands versus 4 bands), can be used for more detailed mapping of floodplains and flood-prone areas on larger scale maps of 1:50,000 or greater. TM data have been used for discriminating land cover classification (Kerber etal., 1985) and to provide useful input to flood forecasting and flood damage models for urban and agricultural areas (Gervin etal., 1985).

Table: *Satellite Imagery Applied To Natural Hazard Assessments*

	EARTHQUAKES	***VOLCANIC ERUPTION***	***LANDSLIDES***	***TSUNAMI***	***DESERTIFICATION***	***FLOODS***	***HURRICANES***
INFORMATION TO BE OBTAINED	Land-use maps, geological maps	Maps of areas vulnerable to lava flows, ash fall, debris fall and fires	Slope maps, slopes stability, elevation, geology, soil type, areas of standing water, land-use maps	Bathymetric/ topographic maps	Land-use maps, soil moisture content, crop condition and natural vegetation	Floodplain delineation on maps, land-use classification, historical data, soil cover and soil moisture	Land-use maps
SPECTRAL BAND	Visible and near IR	Visible, near IR and thermal IR	Visible	Visible, including blue and - near IR	Visible, near IR, and micro wave	Near IR, thermal IR and microwave	Visible and near IR
SPATIAL RESOLUTION	20-80m	30-80m	10-30m	30m	80m-1 km	20m (for cultural features) ; 30-80m (for land use); 1 km (for snow cover and soil moisture)	20m (for cultural features); 30-80m (for land use)
AREAL COVERAGE	Large area	Large area	Large area	Large coastal area	Large regional area	Large regional area	Large area
ALL WEATHER CAPABILITY	No	No	No	No	No	No	No
SYNOPTIC VIEW	Yes	Yes	Yes	Yes	Yes	Yes	Yes
STEREO CAPABILITY	Yes	Yes	Yes	Yes	No	Yes	No
FREQUENCY OF OBSERVATIONS FOR PLANNING STUDY USE	1 to 5 years	1 to 5 years	1 to 5 years	1 to 5 years	Monthly	Seasonal (except weekly for snow cover and soil moisture)	Yearly

Source: Adapted from Richards, PUB. The Utility of Landsat-D and other Satellite Imaging Systems in Disaster Management (Washington, DC: Naval Research Laboratory, 1986).

This approach to floodplain delineation does have its limitations. The area of potential flooding delineated in this manner may represent a flood level that exceeds an acceptable degree of loss. Also, no floods may have occurred during the period of the sensor operation. In this case, indirect indicators of flood susceptibility are used. There are large parts of tropical humid ecosystems where adequate Landsat or other similar imagery is not available due to cloud coverage or heavy haze. In some instances the heavy tropical vegetation masks many geomorphic features so obvious in drier climates. In this case the use of available radar imagery from space or previously acquired from an aircraft survey is desirable. The radar imagery, which has a resolution comparable to Landsat TM and SPOT from both space and sub-orbital altitudes, can satisfactorily penetrate the clouded sky and define many floodplain features. Moisture on the ground noticeably affects the radar return and, together with the textural variations emphasized by the sensor, makes radar a potentially desirable tool for flood and floodplain mapping.

Hurricanes

To mitigate the impact of hurricanes, the planner needs to know the frequency of storms of given intensity in the study area, to what extent these storms could affect people and structures, and what sub-areas would be most affected such as low-lying coastal, estuarine, and reverie areas threatened by flooding and storm surge.

The determination of past hurricane paths for the region can be derived from remote sensing data from the U.S. National Oceanographic and Atmospheric Administration (NOAA) satellite sensors designed and operated for meteorological purposes. These data are already plotted by meteorological organizations in the U.S.A. and other countries where hurricanes are a threat. For plotting new data, the best sensor is the AVHRR which, with its 2,700km swath, makes coverage twice a day, and has appropriate resolution. The red band is useful for defining daytime clouds and vegetation, while the thermal IR band (10.50 mm to 11.50 mm) is useful for both daytime and night-time cloud observations.

Landsat Floodplain Indicators

- Upland physiography
- Watershed characteristics, such as shape, drainage, and density
- Degree of abandonment of natural levees
- Occurrence of stabilized sand dunes on river terraces
- Channel configuration and fluvial geomorphic characteristics
- Backswamp areas
- Soil-moisture availability (also a short term indicator of flood susceptibility)
- Variations in soil characteristics
- Variations in vegetation characteristics

- Land use boundaries
- Flood alleviation measures for agricultural development on the floodplain

Source: Adapted from Rango, A. and Anderson, A.T. "Flood Hazard Studies in the Mississippi River Basin Using Remote Sensing" in Water Resources Bulletin, vol. 10, 1974.

The AVHRR is not useful in other aspects of hurricane contingency planning due to its limited spatial resolution. These planning needs require higher resolution available from other satellite sensors. If imagery of areas inundated by floods, hurricane storms, or other storms is obtained with any sensor immediately after the event, it should be used, of course, regardless of its resolution. Any such information that is obtained in a timely fashion should be used to delineate the problem areas since their definition is more exact than can be interpreted from higher resolution data obtained during a non-flood period.

Predicting areas of potential inundation along coasts and inland can be achieved using topographic maps with scales as large as 1:12,500. When such maps are not available, remote sensing techniques can be used. In areas with a distinct wet and dry season, it is desirable to obtain information for the wet season from Landsat or comparable imagery in the near IR bands, or use a colour IR composite made from Landsat MSS or TM imagery or from SPOT HRV imagery. These image products can be used to identify the moisture-saturated areas susceptible to flooding as well as the higher and drier ground for potential evacuation areas. Likewise, consideration of development plans in view of this potential natural hazard can proceed in a way similar to that for areas prone to flood hazards. For flood hazard assessments, radar imagery from space or aircraft could be used (if available) in lieu of the Landsat MSS imagery. Since there is a general lack of relief in low-lying coastal areas and estuarine areas, stereoscopy would not normally play an important role in this situation. However, stereoscopic viewing even without significant relief enhancement can still reinforce the details of the scene, although at considerably greater cost.

The development planner also needs to consider the additional feature of a hurricane-its high winds. In identifying measures to mitigate wind effects, the planner may consider the type of crops grown, if an agricultural development is being planned, and the design and construction materials used in buildings.

Earthquakes

The planning of development in earthquake-prone areas is laden with problems. There are large human settlements already located in earthquake/prone areas. As with other geologic hazards, the frequency of occurrence can fall in cycles of decades or centuries. Earthquakes are particularly difficult to predict at this time. Thus, mitigation emphasis is on land use planning (non-intensive uses in most hazardous areas), on building strength and integrity, on response planning, and on incorporating mitigation measures into reconstruction efforts. The main problem is the identification of the earthquake damage-prone zones. While in most areas of great earthquake activity some seismic information is available, it may not be sufficient for planning purposes. Remote sensing techniques and resulting data interpretation can play a role in providing additional information.

Tectonic activity is the main cause of destructive earthquakes, followed by earthquakes associated with volcanic activity. Where the history of earthquakes due to seismic activity is present in an area, the faults associated with the activity can frequently be identified on satellite imagery. Where volcanic-related earthquakes occur, the source is generally not as obvious: it may be due to movement on a fault near the surface or deep within the earth, to caldera collapse, or to magma movement within the volcanic conduit.

In order to identify earthquake hazards it is necessary to have the expertise to recognize them and then determine the correct remote sensing tools to best delimit them. Landsat imagery has been effectively and widely used for this purpose since it is less expensive and more readily available than other remote sensing data. Airborne radar mosaics have been successfully used for the delineation of fault zones. Generally, two mosaics can be made of an area: one with the far range portion of the SLAR and the other with the near range portion. The former is best used in areas of low relief where the relief needs to be enhanced, and the latter in areas of high relief where the shadow effect is not needed or may be detrimental to the image.

Radar is applicable to delineate unconsolidated deposits sitting on fault zones-upon which most of the destruction occurs-to identify areas where an earthquake can trigger landslides. This is best accomplished on stereomodels using adjoining and overlapping radar flight lines. Conventional aerial photography, in black and white or colour, would also work well for this purpose.

An alternative, which is adequate but not as good as using radar and aerial photography, is to use multispectral imagery obtained from the Landsat TM and/or MSS or SPOT HRV sensors. Colour IR composites or straight near-IR imagery from these sensors at scales up to about 1:100,000 can be used to define active surface fault zones, but not as efficiently as with the radar images. The distinction between bedrock versus unconsolidated materials and the areas of potential landslide hazards can be defined but, again, only if stereocoverage is available. SPOT sensors can provide this capability.

While radar imagery is an ideal data source, available coverage is extremely limited, and contracting airborne radar is usually prohibitively expensive. Landsat TM and MSS are the most practical data source, simply because of its availability, and both provide sufficient resolution for regional planning studies.

Volcanic Eruptions and Related Hazards

Many hazards are associated with the conditions brought about by volcanic activity. Active volcanoes pose hazards which include the immediate release of expelled ash, lava, pyroclastic flows, and/or poisonous hot gases; volcanic earthquakes; and the danger of mudflows and floods resulting from the rapid melting of snow and ice surrounding the vent during eruption. Some secondary hazards may threaten during volcanic activity or during periods of dormancy. These include landslides due to unstable accumulations of tephra, which may be triggered by heavy rains or by earthquakes. A more detailed discussion of volcanic hazards and their assessment can be found in Chapter 11.

Each volcano has its own particular behaviour within a framework of given magmatic and tectonic settings. Prediction of a volcano's behaviour is extremely difficult, and the best evidence for the frequency of activity and its severity is the recorded history of eruptions. Imminent eruptions are now best recognized by on-site seismic monitoring. Some classifications distinguish between active, inactive, dormant, and extinct volcanoes. But since some of the most catastrophic eruptions have come from "extinct" volcanoes, many volcanologists have abandoned such a classification, settling for a simple distinction between short-term and long-term periodicity.

Gawarecki etal. (1965) first detected volcanic heat from satellite remote sensing using thermal IR imagery from the high resolution IR radiometre (HRIR). Remote sensing data interpretation can lead to the recognition of past catastrophic events associated with recently active volcanoes (recently in the geologic sense), as in the Andes and the Lesser Antilles. This information together with the available historical data can be used as the basis of assessing the risks of an area with potential volcano-related hazards.

The varied nature and sizes of volcanic hazards require the use of various types of sensors from both satellites and aircraft. The relatively small area involved with volcanoes should encourage the use of aerial photography in their analysis. Panchromatic black and white stereo aerial coverage at scales between 1:25,000 to 1:60,000 is usually adequate to recognize and map geomorphic evidence of recent activity and associated hazards. Colour and colour IR photography may be useful in determining the possible effects of volcanic activity on nearby vegetation, but the slower film speed, lower resolution, and high cost diminish much of any advantage they provide. The airborne thermal IR scanner is probably the most valuable tool in surveying the geothermal state of a volcano. The heat within a volcano and underlying it and its movement are amenable to detection. Because of the rapid decrease in resolution with increasing altitude (about 2m per 1,000m) the surveys need to be made at low altitudes under 2,000m.

An IR pattern of geothermal heat in the vicinity of a volcano is an indication of thermal activity which many inactive volcanoes display. Many volcanoes thought to be extinct may have to be reclassified if aerial IR surveys discover any abnormally high IR emissions from either the summit craters or the flanks. Changes in thermal patterns can be obtained for a volcano only through periodic aerial IR surveys taken under similar conditions of data acquisition. The temperature and gas emission changes, however, can be monitored on the ground at ideal locations identified on the thermal imagery, making periodic overflights unnecessary. Continuous electronic monitoring of these stations is possible by transmission through a geostationary data relay satellite, another phase of remote sensing. The thermal IR bands of the satellite sensors now available have inadequate spatial and thermal resolution to be of any significant value to detect the dynamic change in volcanic geothermal activity. In addition to sensing geothermal heat, however, other remote sensing techniques are useful in preparing volcanic hazard zonation maps and in mitigating volcanic hazards. Mitigation techniques requiring photo interpretation and topographic maps include predicting the path of potential mudflows or lava flows and restricting development in those areas.

Landslides

Landslides, or mass movements of rock and unconsolidated materials such as soil, mud, and volcanic debris, are much more common than is generally perceived by the public. Many are aware of the catastrophic landslides, but few are aware that small slides are of continuous concern to those involved in the design and construction business. These professionals can often exacerbate the problem of landsliding through poor planning, design, or construction practices. Frequently, the engineer and builder are also forced into difficult construction or development situations as a result of ignoring the potential landslide hazard. This can be avoided if there is early recognition of the hazard and there is effective consultation between planners and the construction team prior to detailed development planning.

The mass movement of bed rock and unconsolidated materials results in different types of slides, magnitudes, and rates of movement. An area with a potential landslide hazard usually has some evidence of previous occurrences, if not some historical record. Unfortunately, some types of landslides, particularly those of small size, cannot be delineated on remote sensing imagery or through aerial photography. Usually the scars of the larger slides are evident and, although the smaller slide features may not be individually discerned, the overall rough appearance of a particular slope can suggest that mass movement occurred. If a fairly accurate geologic map is available at a reasonable scale (1:50,000 or larger) rock types and/or formations susceptible to landslides may be examined for evidence of movement. An example of this would be finding a shale in a steeper than usual slope environment, implying the strong possibility of a landslide history. An examination of stream traces frequently shows deflections of the bed course due to landslides. If one can separate out the tectonically controlled stream segments, those deflections due to slides or slumps often become evident.

Typical features that signify the occurrence of landslides include chaotic blocks of bedrock whose only source appears to be upslope; crescentic scarps or scars whose horns point downward on a normal-looking slope; abnormal bulges with disturbed vegetation at the base of the slope; large intact beds of competent sedimentary or other layered rock displaced down dip with no obvious tectonic relationship; and mudflow tongues stretching outward from the base of an obviously eroded scar of relatively unconsolidated material. A good understanding of the structural geology of the study area frequently places these superficial anomalies into perspective. As discussed in Chapter 10, the susceptibility to landslides is relative to the area. Landslides can occur on gentle slopes as well as on steep slopes, depending on landscape characteristics.

Most landslide discussions do not address the problem of sinkholes, which are a form of circular collapse landslides. The karstic areas in which they occur are easy to identify even on some satellite imagery

(MSS, TM, SPOT, etc.) due to their pitted appearance and evidence of the essentially internal drainage. Despite the obvious occurrence of many sinkholes, many individual small sinkholes are subtle and not easily recognized. These are frequently the sites of collapse and subsequent damage to any overlying structure when ground water is removed to satisfy development needs, which results in lowering the water table and undermining the stability of the land.

The spatial resolution required for the recognition of most large landslide features is about 10m (Richards, 1982). However, the recognition depends to a great extent on the ability and experience of the interpreter and is enhanced by the availability of stereoscopic coverage, which can be expensive to acquire. Stereoscopic coverage and the resolution requirements preclude use of most satellite-borne sensor imagery, although large block landslides can be detected on Landsat MSS and TM imagery.

Given the spatial resolution requirement, SPOT HRV-P (panchromatic mode) imagery can be useful with its 10m resolution. Its broad band coverage, however, is not conducive to providing adequate contrast in scenes involving heavily vegetated tropics, where most of the potential hazards occur. Ameliorating this factor slightly is the availability of stereocoverage. It is important to understand that this is specifically programmed for the SPOT satellite and that stereocoverage is not normally acquired during sensor operation.

Detection of landslide features is more easily achieved using airborne sensors. Aerial photography with its normal stereoscopic coverage is the best sensor system with which to define landslides, both large and small. Aerial photographic scales as small as 1:60,000 can be used. Black and white panchromatic or IR films are adequate in most cases, but colour IR may prove better in some instances. The IR-sensitive emulsions, as stated earlier, eliminate much of the haze found in the humid tropics. The open water or other moisture in back of recent slump features stands out as an anomaly in the aerial IR stereomodel, either in black and white or colour. The colour IR photography might, in some rare cases, show the stress on the vegetation caused by recent movement. If the scales are large enough, tree deformation caused by progressive tilting of the slope of the soil might also be detected.

A more sensitive detector of moisture associated with landslides is the thermal IR scanner. This sensor is particularly useful in locating seepage areas that lubricate slides. It is particularly effective during the night, when there is a maximum temperature difference between the terrain and the effluent ground water. Despite its utility many factors rule out the widespread use of the thermal IR scanner. These factors include the low altitude required for reasonable spatial resolution, the large number of flight lines required for the large area involved, and the geometrical distortions Inherent in the system. If the terrain to be interpreted has some relief and is nondescript, these distortions become an even greater problem when the data are interpreted by making the location of features very difficult.

SLAR, especially the X-band synthetic aperture radar with its nominal 10m resolution, can be marginally useful in a stereo mode because of its ability to define some larger textures related to landslides. In some cloud-prone environments radar may be the only sensor that can provide interpretable information.

Desertification

Desertification occurs when an ecosystem experiences a diminution or loss of productivity. This process can have a natural and an anthropic component, which may reinforce each other, creating a synergetic effect. The degree of desertification risk is directly related to certain natural conditions such as climate, topography, natural vegetation, soil, and hydrology, as well as to the intensity and type of anthropic activity in the area. Desertification is among the most serious problems of the region. This trend indicates the increasing need to consider desertification processes in integrated development planning studies. Remote sensing, both spaceborne and airborne, provides valuable tools for evaluating areas subject to desertification. Film transparencies, photographs, and digital data can be used for the purpose of locating, assessing, and monitoring deterioration of natural conditions in a given area. Information about these conditions can be obtained from direct measurements or inferred from indicators (keys to the recognition of a desertification process).

In order to describe, evaluate, and decide about the type of action to be taken, the following issues should be addressed:

- Location: involves the identification of areas that are currently undergoing desertification and areas expected to be exposed to the forces that can lead to deterioration.
- Assessment: involves the identification and quantification of vegetative cover types, soils, land forms, and land-use change patterns. Vulnerability to change, rate of change, and direction

of change in desertification patterns can be studied through this assessment.

- Monitoring: accomplished by detecting and measuring changes in landscape characteristics over a period of time. Comparisons are made between present conditions and previously observed conditions for the purpose of recording the reduction in biological productivity.

Chapter 9 presents an initial assessment technique utilizing information commonly available in the early integrated development planning stages. For a more detailed approach, four sets of data should be taken into consideration for a desertification study of a given area: a set taken at the end of the humid season, a set taken at the end of the last dry season, and the same two seasons taken five or ten years earlier (López Ocaña, 1989). Data selection for a given area will be directly related to the desired amount of detail, size of the area, required degree of precision and accuracy, and available time frame.

Large-scale aerial photography provides a great amount of detail for this type of study. Systematic reconnaissance flights can be used for environmental monitoring and resource assessment. Radar sensors and infrared scanners may be used to monitor soil moisture and other desertification indicators. However, acquisition of this type of data is costly and time consuming.

The use of satellite imagery is recommended during the first stages of a detailed desertification study since it offers an overview of the entire region. Computer enhancements, false colour composites, and classifications can offer useful information. Optical enhancements can be performed, but these lack the quantitative control available through an automated approach. Statistical data obtained from a quantitative analysis through the use of a geographic information system can be expressed as a histogram, a graph, a table, or a new image.

AVHRR imagery is commercially available and has been used for vegetation change studies. Ground resolution of 1 to 4 km represents some limitation in making large continental area studies. Other studies have used Nimbus data to delineate moisture patterns and vegetation boundaries. Geostationary Operational Environmental Satellite (GOES) data have been used effectively to locate and measure dust plumes; and Seasat SAP imagery has been applied in the delineation of large dune morphology.

Landsat MSS and TM and SPOT data have proven to be useful and cost effective for regional assessments. Landsat transparencies of bands 5 and 7 have been used to monitor superficial changes in areas undergoing desertification, and to map present water bodies and former drainage systems. Temporal tonal variations on Landsat MSS have been correlated with variations on the field. Movement of sand-dune belts has been detected using Landsat with a multitemporal approach. Albedo changes in arid terrains have been calculated using Landsat digital data: phenomena that tend to lower productivity (increased erosion, loss of vegetation density, deposition of colic sedimentation) also tend to appear brighter on the image. On the contrary, phenomena that tend to increase productivity (increased vegetation, soil moisture), tend to darken the land. In this way, brightness variations can be detected in an area over a period of time. These data can also be calibrated with ground data collected from the areas where change has occurred.

Aerial and space remote sensing provide valuable tools for desertification studies, although, as for any other natural hazard related study, they must be combined with ground-collected data. The use of remote sensing methods should minimize the need for ground data, therefore saving time and resulting quite inexpensive per unit of data. The combination of remotely sensed and ground-collected data can then provide the basis for the assessment.

Bibliography

Auf der Heide, E.: *Disaster Response: Principles of Preparedness and Coordination,* St. Louis: CV Mosby Co., 1989.

Barker, Daniel S.: *Igneous Rocks*, Prentice-Hall, Delhi, 1983.

Batterbury, S.P.J. & A.Warren: *International Encyclopædia of the Social and Behavioral Sciences*, Elsevier Press, U.K., 2001.

Berlin, G.L.: *Earthquakes and the Urban Environment,* CRC Press, Boca Raton, FL, 1980.

Bullard, Fred M.: *Volcanoes of the Earth,* University of Texas Press, Texas, 1984.

Christian, Spencer and Antonia Felix: *Shake, Rattle and Roll: The World's Most Amazing Volcanoes, Earthquakes, and Other Forces*, J. Wiley and Sons, New York, 1997.

Cotton, C.A.: *Volcanoes as Landscape Forms*, Hafner Publishing. New York, 1999.

Dasgupta, Rajdeep: *Disaster Management and Rehabilitation*, Mittal Pub, Delhi, 2007.

Davis, Mike: *Ecology of Fear*, New York, Metropolitan Books, 1998.

Dufka, Corrine: *The Mexico City Earthquake Disaster, Social Casework: The Journal of Contemporary Social Work*, 1988.

Enderle, Judith A.: *Francis, the Earthquake Dog,* Chronicle Books, San Francisco, CA, 1996.

Fedotov, S.A. & Masurenkov: *Active Volcanoes of Kamchatka*, USSR Academy of Sciences/ Nauka Publishers, Moscow, 1991.

Francis, Peter; Oppenheimer, Clive: *Volcanoes,* Oxford University Press, Delhi, 2003.

Freeman, J.R.: *Earthquake Damage and Earthquake Insurance: Studies of a Rational Basis for Earthquake Insurance, also Studies of Engineering Data for Earthquake-resisting Construction*, McGraw-Hill Books, New York, 1932.

Gubbins, David: *Seismology and Plate Tectonics*, Cambridge University Press, North America, 1990.

Hays, J.N.: *Epidemics and Pandemics: Their Impacts on Human History*, Santa Barbara, Calif.: ABC-CLIO, 2005.

Hodge, Paul: *Higher than Everest: An Adventurer's Guide to the Solar System*, Cambridge University Press, UK, 2001.

James, C.D. & Fatemi, S.: *Photographs of the 1906 San Francisco and 1923 Tokyo Earthquakes*, University of California, Berkeley, 2002.

Keyes K.E.: *Emergency Management for Records and Information Programs*, ARMA International, Kansas, 1997.

McBirney, Alexander R. and Williams, Howel: *Volcanic History of Nicaragua*, University of California Press, Berkeley, CA, 1965.

Norris, Debra Hess: *Disaster Recovery: Salvaging Photograph Collections,* Philadelphia, PA, 1998.

Odum, Howard T.: *Environment, Power and Society*, New York, Wiley-Interscience, 1971.

Pettigrew, Mark: *Planet Earth,* Price Stern Sloan, Los Angeles, CA, 1987.

Reiter, L.: *Earthquake Hazard Analysis: Issues and Insights*, Columbia University Press, New York, 1990.

Schmincke, Hans-Ulrich: *Volcanism*, Springer-Verlag., Delhi, 2004.

Sigurdsson, Haraldur: *Melting the Earth: The History of Ideas on Volcanic Eruptions*, Oxford University Press, Delhi, 1999.

Singh, Jagbir: *Disaster Management: Future Challenges and Opportunities*, I K Pub, Delhi, 2007.

Thordarson, Thor; Hoskuldsson, Armann: *Iceland,* Terra Publishing, UK, 2002.

Wahrhaftig, Clyde: *A Streetcar to Subduction: and Other Plate Tectonic Trips by Public Transportation in San Francisco*, American Geophysical Union, Washington, D.C., 1984.

Williams, Howel: *Volcanic History of the Guatemalan Highlands*, University of California Press, Berkeley, CA, 1960.

Worster, Donald: *Dust Bowl: The Southern Plains in the 1930s*, New York, Oxford University Press, 1979.

Zenaida, D.: *Women and Children During Disaster: Vulnerabilities and Capacities*, Elaine Enarson and Betty Hearn Morrow, 1995.

Index

M

N

P

R

S

T

V

W

❑❑❑